Solid-State Electronic Devices

Solid-State Electronic Devices

Editor

Beniamino Cipriani

scitus academics

Soil Mechanics Fundamentals

Edited by **Beniamino Cipriani**

Printed in 2017

ISBN: 978-1-68117-173-9
Library of Congress Control Number: 2015951999

SCITUS Academics LLC,
616, Corporate Way, Suite 2, 4766,
Valley Cottage, NY 10989

www.scitusacademics.com

Notice

Preface

This book provides a modern and concise treatment of the solid state electronic devices that are fundamental to electronic systems and information technology. Solid state electronic devices are those circuits or devices built completely from solid materials and in which the electrons, or other charge carriers, are kept entirely within the solid material. The term is often used to contrast with the earlier technologies of vacuum and gas-discharge tube devices, and it is also conventional to exclude electro-mechanical devices from the term solid state. While solid-state can include crystalline, polycrystalline and amorphous solids and refer to electrical conductors, insulators and semiconductors, the building material is most often a crystalline semiconductor. The main devices that comprise semiconductor integrated circuits are covered in a clear manner accessible to the wide range of scientific and engineering disciplines that are impacted by this technology.

This book can expect to derive a solid foundation for understanding modern electronic devices and also be prepared for future developments and advancements in this far-reaching area of science and technology.It presents basic and state-of-the-art topics on materials physics, device physics, and basic circuit building blocks which will be useful to researchers as well as practicing engineers.

Table of Contents

CHAPTER 1 Recent Progress on Thin-Film Encapsulation Technologies for Organic Electronic Devices 1

ABSTRACT 1
BACKGROUND OF ENCAPSULATION TECHNOLOGY FOR ORGANIC ELECTRONICS 2
DEGRADATION MECHANISM OF ORGANIC ELECTRONIC DEVICES 3
MEASUREMENT OF PERMEATION RATE 4
TRADITIONAL ENCAPSULATION METHODS 7
CHEMICAL VAPOR DEPOSITION TECHNOLOGY 10
ALD TECHNOLOGY 11
Figure options 14
THIN FILM ENCAPSULATION FOR FLEXIBLE DEVICES 15
SUMMARY 18
ACKNOWLEDGEMENTS 18
REFERENCES 19
CITATION 23

CHAPTER 2 A Quantum Energy Transport Model for Semiconductor Device Simulation 25

ABSTRACT 25
INTRODUCTION 25
4 MOMENTS QUANTUM ENERGY TRANSPORT MODEL 26
DISCRETIZATION AND ITERATIVE SOLUTION METHOD 34

Discretization 34
Iterative solution method 37
NUMERICAL RESULTS 41
CONCLUSION 47
ACKNOWLEDGEMENT 47
REFERENCES 47
CITATION 49

CHAPTER 3 Modeling and Simulation of Dynamics and Noise of Semiconductor Lasers Under Ntsc Modulation for Use in the Catv Technology 51

ABSTRACT 51
INTRODUCTION 52
THEORETICAL AND CALCULATION MODEL 56
RESULT AND DISCUSSIONS 58
Light-current (L-I) characteristics 58
Small-signal modulation transfer function 59
Characterization of the modulated laser signal 61
(Modulation index versus channel frequency) diagram of modulation dynamics 65
Noise characteristics of the modulated types of the SL signal 66
CONCLUSIONS 67
REFERENCES 68
CITATION 71

CHAPTER 4 Advances in Atomic Physics: Four Decades of Contribution of the Cairo University – Atomic Physics Group 73

ABSTRACT 73
INTRODUCTION 73
ELECTRON–ATOM COLLISIONS 74
The first table-top synchrotron 76
ION–ATOM COLLISIONS 81
ATOMIC STRUCTURE CALCULATIONS AND X-RAY LASERS 88
LASER-INDUCED BREAKDOWN SPECTROSCOPY (LIBS) 94
LASER COOLING AND BOSE–EINSTEIN CONDENSATION 103
Conflict of interest 113

Compliance with Ethics Requirements....113
REFERENCES....113
CITATION....124
CHAPTER 5 A Modern Perspective on the History of Semiconductor Nitride Blue Light Sources....125

ABSTRACT....125
HIGHLIGHTS....126
EARLY HISTORY OF SEMICONDUCTOR LIGHT EMITTERS....126
NITRIDE SEMICONDUCTORS....127
GALLIUM NITRIDE LIGHT-EMITTING DIODES....130
Mg-DOPED BLUE GaN LED....134
RETURN TO RCA LABS....137
WORLD LOSES INTEREST IN NITRIDES....138
RENEWED INTEREST AND MAJOR PROGRESS....140
BUT WHY WERE THEY BLUE?....144
TRUE BLUE NITRIDE LIGHT-EMITTING DEVICES....146
AFTERWORD....152
REFERENCES....153
CITATION....155

CHAPTER 6 **PNP Pin Bipolar Phototransistors for High-Speed Applications Built in a 180 Nm CMOS Process........** Error! Bookmark not defined.**57**

ABSTRACT....157
HIGHLIGHTS....158
INTRODUCTION....158
WORKING PRINCIPLE....161
METHODOLOGY....162
IMPLEMENTED PHOTOTRANSISTORS....165
RESULTS AND DISCUSSION....166
Electric field strength and space-charge region simulation....167
DC characterization....169
AC characterization: responsivity and bandwidth....173
CONCLUSION....179
ACKNOWLEDGEMENTS....180
REFERENCES....181
CITATION....182

CHAPTER 7 **Subband Engineering in *N*-Type Silicon Nanowires Using Strain and Confinement....183**

ABSTRACT....183
HIGHLIGHTS....184
INTRODUCTION AND MOTIVATION....184
SUBBAND STRUCTURE MODELING....185
Bulk Hamiltonian....185
Quantization of the Hamiltonian....187
Stress and strain....188
Discretization and numerical solution....189
Group velocity and effective mass - postprocessing....191
RESULTS AND DISCUSSION....194
Stress behavior of [1 1 0] and [1 1 1] nanowires....195
Influence of the aspect ratio on stress behavior....200
Comparison of nanowires with thin films....203
SUMMARY AND CONCLUSIONS....204
ACKNOWLEDGMENT....205
REFERENCES....205
CITATION....207

CHAPTER 8 **The Different Roles of Charged and Neutral Atomic and Molecular Oxidising Species in Silicon Oxidation from AB Initio Calculations....209**

ABSTRACT....209
INTRODUCTION....210
ATOMIC-SCALE PROCESSES IN SILICON OXIDATION AND BASIC MODELS....211
TECHNIQUES....214
RESULTS FOR OXIDISING SPECIES IN QUARTZ....216
OXIDISING SPECIES IN AMORPHOUS SiO_2....221
ROLES OF ATOMIC AND MOLECULAR SPECIES IN OXIDATION AND THE CONSEQUENCES OF CHARGING....222
ACKNOWLEDGEMENTS....223
REFERENCES....223
CITATION....225

CHAPTER 9 Strained Mosfets on Ordered Sige Dots 227

ABSTRACT...227
INTRODUCTION...228
GROWTH OF REGULAR ARRAYS OF SIGE ISLANDS..........................228
MODELING OF THE STRAIN FIELD IN THE DOTFET PROCESS 230
THREE-DIMENSIONAL DEVICE SIMULATION....................................232
Structure definition...233
Electrical characterization...235
Three-dimensional impact of the dot 237
Comparison with an SOI structure...238
DISCUSSION 240
CONCLUSION 241
ACKNOWLEDGMENT 241
REFERENCES 242
CITATION...243

CHAPTER 10 Impact of Oxidation and Reduction Annealing on the Electrical Properties of Ge/La_2O_3/ZrO_2 Gate Stacks 245

ABSTRACT...245
HIGHLIGHTS..246
INTRODUCTION...246
EXPERIMENTAL 248
Formation of gate stacks 248
Electrical and structural characterization of MOS capacitor structures 249
RESULTS 250
Capacitance–voltage characterization...250
Interface trap density and equivalent oxide thickness 252
Structural characterization 254
DISCUSSION 257
CONCLUSION 259
ACKNOWLEDGMENTS 260
REFERENCES 260
CITATION...262

CHAPTER 11 **Automatic Generation Control with Thyristor Controlled Series Compensator Including Superconducting Magnetic Energy Storage Units..... 263**

ABSTRACT.....263
INTRODUCTION.....264
MATERIALS AND METHODS.....266
System under study.....266
Control structure and objective function.....267
Modeling of TCSC in AGC.....269
Modeling of SMES in AGC.....272
Over view of differential evolution.....273
Initialization of parameter.....274
Mutation operation.....274
Crossover operation.....274
Selection operation.....275
RESULTS AND DISCUSSIONS.....275
Implementation of DE.....275
Analysis of results.....278
Sensitivity analysis.....283
CONCLUSION.....290
APPENDIX A.....290
REFERENCES.....291
CITATION.....293

Index.....295

CHAPTER 1

Recent Progress on Thin-Film Encapsulation Technologies for Organic Electronic Devices

Duan Yu, Yong-Qiang Yang, Zheng Chen, Ye Tao, Yun-Fei Liu

State Key Laboratory on Integrated Optoelectronics, College of Electronic Science & Engineering, Jilin University, Changchun 130012, PR China

ABSTRACT

Among the advanced electronic devices, flexible organic electronic devices with rapid development are the most promising technologies to customers and industries. Organic thin films accommodate low-cost fabrication and can exploit diverse molecules in inexpensive plastic light emitting diodes, plastic solar cells, and even plastic lasers. These properties may ultimately enable organic materials for practical applications in industry. However, the stability of organic electronic devices still remains a big challenge, because of the difficulty in fabricating commercial products with flexibility. These organic materials can be protected using substrates and barriers such as glass and metal; however, this results in a rigid device and does not satisfy the applications demanding flexible devices. Plastic substrates and transparent flexible encapsulation barriers are other possible alternatives; however, these offer little protection to oxygen and water, thus rapidly degrading the devices. Thin-film encapsulation (TFE) technology is most effective in preventing water vapor and oxygen permeation into the flexible devices. Because of these (and other) reasons, there has been an intense interest in developing transparent barrier materials with much lower permeabilities, and their market is expected to reach over $550 million by 2025. In this study, the degradation mechanism of organic electronic devices is reviewed. To increase the stability of devices in air, several TFE technologies were applied to provide efficient barrier performance. In this review, the degradation mechanism of organic electronic devices, permeation rate measurement, traditional encapsulation technologies, and TFE technologies are presented.

BACKGROUND OF ENCAPSULATION TECHNOLOGY FOR ORGANIC ELECTRONICS

Organic electronics have been the focus of many investigations in the fields of physics and chemistry for more than 50 years. In the early 1960s, Pope et al. and Helfrich et al. found light transmitted from single crystal of anthracene [1] and [2]. Their study showed the possibility of achieving light-emitting devices from molecular crystals. Conducting polymers invented in 1977 is another important discovery. These novel conducting polymers provide tremendous possibilities for the manipulation of organic electronic devices [3]. Moreover, the Nobel Prize in chemistry in 2000 was shared by H. Shirakawa, A.G. MacDiarmid, and A.J. Heeger for the invention of organic electronic devices. Even though the previous studies showed the potential of electronic devices, the progress in applications of organic materials is still far from expectations. Until 1980s, the interest of researchers in undoped organic semiconductors significantly increased owing to the demonstration of a series of efficient organic electronic devices. A two-layer organic photovoltaic cell with low voltage and efficient thin-film light emitting diode was fabricated by Tang et al. in 1986 and 1987 [4] and [5]. In 1986, the first transistor based on organic semiconductor was reported [6], and this led significant interest of researchers in the field. Compared to silicon-based electronic devices, the efficiency of organic devices such as LEDs, solar cells, and transistors is less. However, the organic electronic devices can be fabricated flexibly and will be the promising candidate for ultimate technology to customers and industries in the near future. However, organic materials are unstable when operating in the ambient conditions. Therefore, achieving flexible devices is a big challenge [7]. In fact, many technologies were used to encapsulate organic electronic devices. The achievement of the companies active in the development of high barrier films is remarkable. A well-known approach called Barix technology can be efficient to protect devices from the corrosion of water vapor and oxygen permeation. This technology involves the deposition of alternative dyads of organic and inorganic films as the barrier layers [8]. Samsung SDC has adopted direct encapsulation by TFE using the Vitex multilayer technology with the implementation of three or less dyads. This technology with variation in the deposition techniques is currently used in the development and production and is expected to hold for the next generation of foldable displays. However, up to now, no mature product by thin-film encapsulation technology is in the market, as it is expensive because of low throughput and high investment. Recently, an alternative technology called atomic layer deposition (ALD) was applied to encapsulate organic devices. The barrier layer fabricated by ALD showed identical barrier performance with thinner barrier layers than other encapsulation methods [9]. However, the total growth time of the deposition is still too long to be applicable in industries.

In this article, the degradation mechanism caused by water vapor and oxygen is discussed. Various barrier performance measurements and encapsulation methods will also be discussed in the later part. After introducing the traditional encapsulation techniques, thin film encapsulation (TFE) methods such as sputter, chemical vapor deposition (CVD) and ALD will be discussed. Finally, the TFE technique on the flexible electronic devices will be presented.

DEGRADATION MECHANISM OF ORGANIC ELECTRONIC DEVICES

Organic light-emitting devices (OLEDs) have undergone rapid development among all the organic electronic devices. The degradation of OLEDs has been reported in literature by various researchers. The degradation of small molecule organic materials, migration of ionic species, electrochemical reactions at the electrode/organic interface, dark spot growth, and the effect of water vapor and oxygen have been reported [10], [11], [12], [13] and [14]. In 1992, Adachi et al. first compared the devices operated at different conditions and showed that the luminance of the devices operated in the ambient condition decreased from 115 cd/m^2 to 1 cd/m^2 during 2.5 h. However, the device worked well for 15 h under vacuum with the same decay trend of luminance, indicating that water vapor and oxygen were the main reason for the decay of OLEDs [15]. By analyzing the degradation mechanisms in OLEDs, Schaer et al. found that water vapor is thousand times more destructive than oxygen at room temperature. In general, dark particles deposited during the fabrication led to the formation of pinholes in the cathode. Water vapor pass through the pinholes and diffuse into the cathode/organic interface, producing hydrogen gas by reacting with the cathode. Under the pressure of the gas, bubbles are formed. Through the bubbles, water can be transported much faster than by diffusion. Eventually, the bubbles burst and give rise to additional entry ports for water vapor [14]. The degradation mechanisms of organic photovoltaic devices (OPV) and organic thin-film transistors (OTFT) are the same to OLEDs [16], [17] and [18]. Therefore, the devices require protective layers with extremely low water vapor and oxygen permeability. Other studies have shown that the water vapor permeation rate should be less than 10^{-6} g/m²/day, and oxygen permeation rate should be less than 10^{-3} cc/m²/day. The minimum values are required to assure adequate lifetime for most OLEDs [8].

MEASUREMENT OF PERMEATION RATE

Until now, the most common test method to evaluate the encapsulation of organic electronic is electrical analysis of calcium (Ca) corrosion [19] and [20], usually called as Ca corrosion test, demonstrated by Paetzold et al. in 2003 [21]. Numerous approaches were adopted before the Ca corrosion test. Tritiated water test, first introduced in 1992 by Coulter et al. [22] improved the accuracy of water vapor transmission rate (WVTR), representing a lower WVTR value below 10^{-6} g/m²/day. The method was used by Hansen et al. and Groner et al. in 2001 and 2006, respectively [23] and [24]. In Groners work, tritiated water used as a radioactive tracer was placed on the bottom flange, resulting in a 100% relative humid environment on the downstream side of the encapsulating film, as shown in Fig. 1.

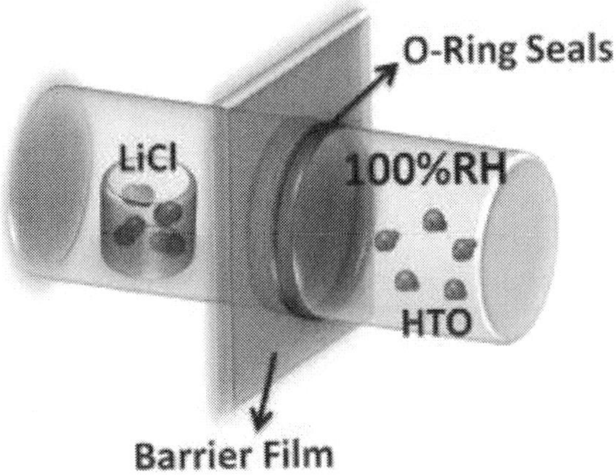

Figure 1. A layout of HTO water test demonstrated in Groners work.

Moreover, a vial containing LiCl was suspended in the top part of the chamber. The container absorbed the tritiated water (HTO) and water permeating through the film. To calculate the HTO transmission rates, the decay of tritium must be counted by a scintillation counter. Cros et al. used controlled relative hygrometry (HDO) to measure the permeation. The HDO test was similar to the HTO test [25], with a favorable detection limit, and the result of the HDO method needed further study because of the fact that the absorption of HTO (HDO) can be affected by the water vapor from the outside atmosphere. Furthermore, the test was dangerous, because of the radioactivity of the radioactive tracer and unavailability in some experimental situations.

The method using a humidity sensor provided by the commercial technology was another measurement method for the permeation of water vapor as well as oxygen, resulting in the oxygen transmission rate (OTR). Instruments developed by MOCON Inc. were used for measuring the WVTR and OTR in numerous studies [23], [26] and [27], supporting the user-friendly operating process. However, the sensitivity of the method, 10^{-3} g/m^2/day, was not sufficient for the measurement of barrier film especially on OLEDs.

Ca, as a conducting and opaque metal, become non-conducting and transparent after the oxidation, making Ca corrosion test possible for measuring the permeation rate through the film and can be calculated by the following equation:

$$P = -n\frac{M\,(H_2O)}{M\,(Ca)}\delta\rho\frac{l}{b}\frac{d(1/R)}{dt} \tag{1}$$

where δ is the density of Ca, l and b are length and width, respectively, of Ca layer, and ρ is the Ca resistivity, indicating that the permeation rate is proportional to differential of the conductance curve $1/R$ versus measurement time t. Ca reacts with water following the reaction shown in the following equations:

$$2Ca+O_2 \rightarrow 2CaO \tag{2}$$

$$Ca+H_2O \rightarrow CaO+H_2 \tag{3}$$

$$Ca+H_2O \rightarrow Ca(OH)_2 \tag{4}$$

The value of n can differ, because each Ca atom can react with half molar equivalent oxygen (n=0.5) molecule or up to two water molecules (n=2) [21]. A WVTR can be obtained from an improved function by the adjustable n [28]

$$\begin{aligned}&\text{WVTR}\,[\text{g/m}^2/\text{day}]\\&= -n \times \delta_{Ca} \times \rho_{Ca} \times \frac{d}{dt}\left(\frac{1}{R}\right) \times \frac{M(H_2O)}{M(Ca)}\frac{Ca\_Area}{Window\_Area}\end{aligned} \tag{5}$$

Moreover, the analysis of the permeation can be either electrical [29] or optical [19]. Both these are highly sensitive, because the oxidation can be monitored by measuring the resistivity of Ca compounds as well as tracking the transparency. Layouts of both the analysis methods are shown in Fig. 2.

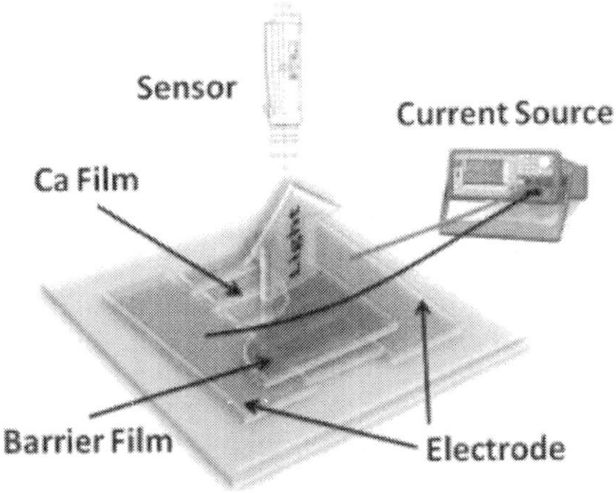

Figure 2. Electrical Ca test analyzed by measuring source and optical Ca test using spectrum sensor.

TRADITIONAL ENCAPSULATION METHODS

The commercial products based on OLEDs such as OLED display, lighting product are available now, and most of displays are based on the solid substrates. The encapsulation method is achieved by glass and metal lid [30]. In 1994, Burrows et al. presented a simple encapsulation technique for OLEDs. The design for the encapsulated OLEDs is shown inFig. 3(a). Glass palates were used as the substrates, and indium tin oxide (ITO) was coated on the substrates. The organic layers and the top electrodes are grown by the thermal evaporation. Then, the device is transferred to a glove box under nitrogen. The epoxy adhesive is introduced around the edge of the device using a syringe. Finally, a clean metal or glass lids covers at the top of the device. An inert nitrogen gas is filled in the sealed volume. The encapsulated devices maintain 40% of the initial luminance after more than 1000 h continuous operation [30]. In addition, desiccant materials such as Ca and barium were used to absorb water vapor diffusing from the epoxy adhesive and removing any existing water [31]. The sealing of organic devices between the two glass substrates was developed using plastic substrates with barrier films [32]. Recently, Yang et al. performed a very simple and convenient method with UV-curable polymer (NOA63 from Norland Optics) film as a passivation layer, acting as a temporary barrier for OLEDs. The NOA63 protective layer significantly restricted the moisture penetrating into the OLEDs, without affecting its performance. The decay time (1860 min) of the device reached 70% of the initial luminance and was 6.8 folds higher than that of the device without encapsulation. The effective water vapor transmission rate was 0.031 $g/m^2/day$ at 20 °C and 50% *RH* [33].

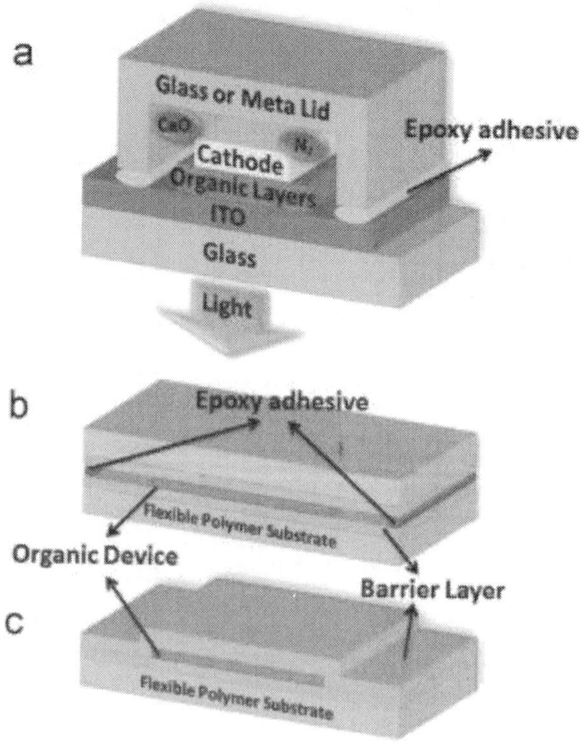

Figure 3. (a) Schematic side view of an encapsulated OLED with traditional encapsulation; (b) coated flexible lid; (c) thin-film.

However, these typical encapsulation techniques with rigid materials have some problems in application in flexible devices [34]. Moreover, flexible encapsulation approaches such as ultra-thin glass, barrier-coated flexible lids, and vacuum-deposited thin films are more effective, as shown in Figs. 3(a)–(c) [35], [36] and [37]. In 1983, Jamieson et al. utilized vacuum deposition to obtain aluminum (Al) layers (10–100 nm) on polyester film, as an attempt of flexible cathode and thin-film defect layer. The Al layers were polycrystalline with the grain size of the order of the metal thickness. The measured oxygen permeability of the metallized films correlated linearly with the observed density of 2–3 μm diameter pinhole defects in the Al coating; however, it has no significant relationship to the coating thickness. The major cause of the pinhole defects is the presence of dust particles on the polymer film surface during the metallization, which subsequently become dislodged and leave an unmetallized shadow. The damage by scuffing of particles can also be a source of pinhole defects, which are often observed as "runs" [38]. In 2005, Jeong et al. investigated the Al cathode generated by ion-beam-assisted deposition (IBAD), which significantly improved the passivation properties than that by the thermal evaporation. The dense and highly packed Al cathode effectively prohibited the permeation of H_2O and O_2 from pinhole defects, slowing the growth of black spots [39].

Attempting metal deposition may provide the possibility of thin film encapsulations, overcoming the barrier layer deposition of metal oxide and molecules. Different methods are applied in manufacturing thin films. Chemical vapor deposition (CVD) as well as atomic layer deposition and molecular layer deposition (ALD/MLD) are most frequently used methods [7], [40] and [41]. Compared to the CVD, the ALD/MLD shows excellent surface properties owing to its self-limit growth mechanism, suffering from extremely slow processes. Moreover, the CVD reactions can be performed at atmospheric pressure, whereas the ALD/MLD processes can only be performed in a low vacuum. Because of there limitations, the surface characteristics of the ALD-processed thin films exhibit much lower permeabilities [42] and [43]. Therefore, both these methods fit for different demands and will be discussed in detail in the following sections.

CHEMICAL VAPOR DEPOSITION TECHNOLOGY

Several chemical vapor deposition methods such as atmospheric pressure chemical vapor deposition (APCVD), plasma enhanced chemical vapor deposition (PECVD), and low pressure chemical vapor deposition (LPCVD) have been developed to deposit thin films (especially SiN_x film) as the encapsulation layer [40]. All of them are inexpensive with mass production. The APCVD technique has a lot of advantages such as good step coverage, uniformity, and relatively low cost. However, this type of deposition process works at high temperatures. Therefore, to lower the deposition temperature, the PECVD was adopted. PECVD is a type of deposition technology, allowing industrial-scale deposition of good quality insulating films such as silicon nitride and silicon oxide with good adhesion [44] and [45]. The most important advantage of PECVD is that it can be performed at low deposition temperatures. Thermal CVD requires deposition temperatures in the range 700–900 °C, whereas only 250 °C or lower temperature is required to deposit similar films by PECVD. Because of the intrinsic properties of organic materials and polymer substrates for flexible electronics devices, the deposition temperature is limited mostly below 100 °C [46]. Therefore, it is common to deposit films by PECVD at low temperatures for organic device fabrication.

In the mid-1970s, PECVD was first used in photovoltaic (PV) devices (amorphous silicon solar cells). In 1981, Hezel and SchoKrner transferred the fabrication of SiN_x by PECVD from the microelectronics industry to the crystalline silicon PV community. After that, metal–insulator–semiconductor inversion layer (MIS-IL) cells were developed by using plasma silicon nitride [47]. In 1999, Erlat et al. deposited SiO_x gas barrier coatings on polymer substrates by PECVD. His study demonstrated that it was possible to use SiO_xor SiN_x thin films in biomedical device applications and food packaging, indicating that thin film deposited by PECVD had good water vapor barrier performance and it can be deposited on polymer surfaces owing to its flexibility [48]. Six years later, Wuu et al. used SiO_x and SiN_x films deposited by PECVD to encapsulate the OLED. Even more, they used multilayered structures of multiple polymer/inorganic layers to improve the barrier performance. In their study, the WVTR and OTR of SiO_x (50 nm)/SiN_x (50 nm) barrier coatings on PC at 80 °C decreased to 0.01 $g/m^2/day$ and 0.1 $cm^3/m^2/day$, respectively. This result caused significant repercussions in the encapsulation technology, broadening the aspects of PECVD encapsulation prospects [43] and [46]. The barrier properties of the PECVD deposited film by numerous studies are summarized in Table 1.

Table 1. Summary of the barrier properties of PECVD thin films including material and barrier structure.

Material	Density	Thickness (nm)	WVTR ($g/m^2/day$)	Transparency (%)	OLED lifetime	Ref.
SiN_x	2.225	100	5×10^{-2}	85	8000	[49]
CF_x/Si_3N_4		100	5×10^{-2}		8000	[20]
SiN_x	2.336	100	4×10^{-2}	85		[50]
SiN_x	1.954	500	3×10^{-2}	90	7500	[51]
SiO_x/SiN_x		100	1×10^{-2}	92	7000	[43]

Optimum film stress, optical transparency, low surface roughness, good mechanical behavior, and low deposition temperature are the requirements for barrier films for flexible OLEDs. The amorphous hydrogenated thin films deposited by PECVD at intermediate substrate temperatures have good adhesion and good coverage of complicated substrate shapes. All the significant and encouraging results make PECVD technique increasingly attractive. The advanced success in PECVD encapsulation will be the need of the future.

ALD TECHNOLOGY

ALD, a technique widely used nowadays, especially in TFE, was developed and introduced worldwide in the late 1970s [7]. When the term atomic layer epitaxy (ALE) was used, the first ALE (nowadays called ALD) was reported in 1977 by Suntola et al. [52], because of its self-limiting surface reactions, ALD is better described in literature. The principles of the ALD method are based on sequential, alternate, self-limiting surfaces reactions on the substrate showing in Fig. 4. The self-limiting growth mechanism ensures the growth of thin film of excellent uniformity and conformity with accurate thickness on large and complex surfaces [53], [54], [55] and [56]. The thin films are grown in an ALD reactor, described in detail in literature [57]. Compact films grown by ALD are achieved by an AB binary sequential reaction, separated by washing flow (mostly N_2). The characteristic shows that ALD as a very promising technique for thin film encapsulation and is an important opportunity for the development for organic electronics industry.

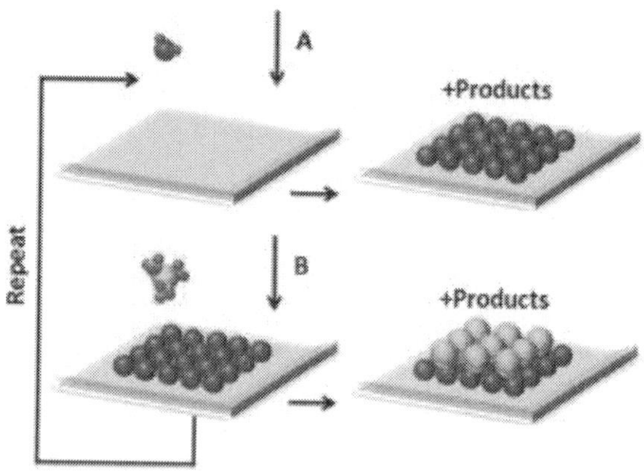

Figure 4. Schematic representation of ALD: AB acted as organometallic precursors sequentially reacting with oxidant.

Al_2O_3 thin film growth at lower deposition temperatures using trimethylaluminum (TMA) and H_2O by ALD is reported. The elemental analysis of the thin film revealed that the hydrogen content increased with decreasing growth temperature. In the meantime, the densities of the Al_2O_3 thin film decreased as the temperature decreased, and the densities were 3.0 and 2.5 g/cm^3 at 177 and at 33 °C, respectively [58]. Remarkably, with decreasing growth temperature, the reaction time became longer, because of the slower reaction rates and longer purge time. Sarkar et al. substituted O_3 for H_2O as the ALD oxidant, and the O_3-based Al_2O_3 film showed superior encapsulation performance compared to the H_2O-based Al_2O_3 film [59] and [60]. In recent years, the comparison between H_2O- and O_3-based Al_2O_3 films has been possible. The barrier characteristics of

O_3-based films are better than the H_2O-based Al_2O_3 films reported by Yang et al., and the TFE showed a lower WVTR value of 8.7×10^{-6} g/m²/day [61]. Hybrid nanolaminates improving the encapsulation performance [9] besides single layer ALD thin films in literature are summarized in Table 2.

Table 2. Summary of the barrier properties of ALD thin films including material and temperature.

Material	Temperature (°C)	Thickness (nm)	WVTR (g/m²/day)	Transparency (%)	OLED lifetime (h)	Ref.
Al_2O_3		100	4.7×10^{-5}	85	8000	[42]
Al_2O_3/TiO_2	100	50		90		[65]
Al_2O_3/ZrO_2		40	3.2×10^{-4}	95		[67]
Al_2O_3/ZrO_2	80	130	4.7×10^{-5}	70	10,000	[64]
Al_2O_3/SiO_2	38	86	5×10^{-5}			[9]

By prolonging the permeation path for water vapor, a lower WVTR value can be obtained. According to some researches [62], [63] and [54], both Al_2O_3 single film and Al_2O_3/ZrO_2 hybrid film exhibited an amorphous structure, whereas ZrO_2 single film showed a monoclinic structure, as shown in Fig. 5. The formation of a more stable $ZrAl_xO_y$ phase with less hydrogen bonding resulted in better barrier properties [63]. The reported WVTR value for 30 nm Al_2O_3/ZrO_2 hybrid film was 2×10^{-4} g/m²/day at 85 °C and 85% *RH* [62]. Moreover, Al_2O_3/TiO_2 (ALD) nanolaminates demonstrated superior encapsulation performances as well. The WVTR of 50 nm hybrid nanolaminate was 1.81×10^{-4} g/m²/day [63]. Aarti et al. presented a hybrid architecture consisted of 10 nm O_3-based Al_2O_3/TiO_2 starting layer with 90 nm H_2O-based Al_2O_3/TiO_2 film. A lower WVTR of 10^{-3} g/m²/day at 38 °C and 90% *RH* was achieved [66].

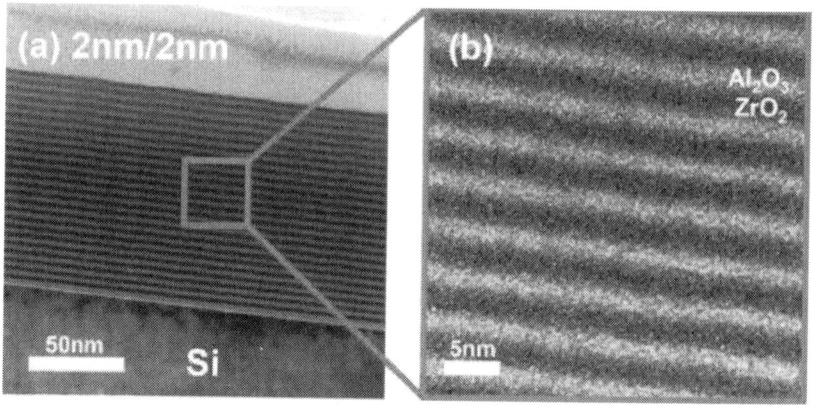

Figure 5. Cross-sectional transmission electron microscopic images of 2 nm/2 nm multi-layer at (a) low resolution and (b) high resolution.

Figure options

Recently, many researchers studied the emerging molecular layer deposition (MLD) technique to fabricate hybrid organic/inorganic thin films, exhibiting excellent mechanical stability and flexibility [41] and [68]. MLD is similar to ALD and can be deposited by the same equipment with different precursors. The deposition of hybrid films with no transfer-induced impurities is very convenient. Sundberg et al. [69] introduced a variety of hybrid organic/inorganic thin films. When organometallic precursors reacted with alcohols or phenols, the resulting films were called metalcone [68]. Many researchers studied the barrier performance of the hybrid organic (MLD)/inorganic (ALD) films [70], [71], [72], [73], [74], [75], [76], [77] and [78]. The WVTR value of Al_2O_3/alucone films was 2.08×10^{-2} g/m²/day, which was lower than that of Al_2O_3 or alucone single layer under the same conditions [72]. The tunable optical characteristics of Al_2O_3/alucone are reported by Sun Feng-Bo et al. and a lower WVTR value of 8.68×10^{-5} g/m²/day was obtained [79].

THIN FILM ENCAPSULATION FOR FLEXIBLE DEVICES

The study for the development of flexible organic light emitting diodes (FOLEDs) is rapidly increasing worldwide, because FOLEDs will radically change several aspects of daily life. In particular, with the exploration of flexible electrodes, flexible polymeric substrates, and flexible encapsulation technique, the low-cost roll-to-roll production processes will be possible to broaden the application areas such as lighting and displaying [80] and [81].

However, some issues about the flexibility of OLEDs have to be solved. Flexible polymer substrates used in FOLEDs are flexible than the traditional glass or metal substrate, but have insufficient barrier performance against water vapor and oxygen. According to literature [31] and [82], the calculated WVTR and OTR values for some different polymer substrates are 10^{-1}–40 g/m^2/day and 10^{-2}–10^2 cm^3/m^2/day, respectively, and were determined by the instinct aspects of the materials. These values are much higher than those for the rigid glass substrate. Moreover, the polymer substrates such as PET, PEN, and PI are rougher than glass or metal substrate, because they cannot be flattened through some mechanical processing tools. The surface roughness is also very important for the performance and lifetime of FOLEDs [83] and [84]. Therefore, a smooth barrier layer on flexible polymer substrate is needed for FOLEDs [7] and [85].

Inorganic/organic hybrid encapsulation film is known to be superior compared to single inorganic layer in the barrier performance in FOLEDs, mainly because of two reasons: (i) the organic layer in nanolaminates functions as a type of resistive interlayer that appears to lengthen the diffusion path for water permeation, (ii) nanolaminates inhibit the propagation of the defects through the multilayer structure. Both these characteristics are beneficial as they improve the associated WVTR values [86] and [87]. In 2002, Weaver et al. fabricated an air-stable organic light-emitting diodes using a 175 mm thick PET substrate coated with an Al_2O_3/polymer multilayered barrier film. The estimated WVTR value through the plastic substrate was 2×10^{-6} g/m^2/day. A lifetime of

3800 h from an initial luminance of 425 cd/m^2 was achieved [32] and [88]. In 2006, a multilayer thin film encapsulation method was adopted by Kang et al. to protect an organic layer from water vapor and oxygen. The hybrid thin films were deposited onto PET by electron beam and sputtering. The SiON/SiO_2 and parylene layer showed the best barrier properties. The WVTR of the PET substrate decreased from 0.57 g/m^2/day (pristine substrate) to 1×10^{-5} g/m^2/day by the application of SiON and SiO_2 layers, showing immense potential for the FOLED applications [89], [90] and [91]. With the development of ALD technique, more and more researchers have focused on the ALD deposited thin films in the encapsulation of FOLEDs. In 2012, Vähä-Nissi et al. first studied the Al_2O_3/alucone hybrid film deposited by ALD and MLD. They found that with alucone (MLD) organic layer insertion, the number and size of defects were smaller compared to the thick brittle Al_2O_3films after straining, thus improving the barrier performance [92]. In 2013, some Korean research groups studied the ALD-based inorganic film combined with different organic films by spin coating on flexible substrate. As shown in Fig. 6, Choi et al. obtained a low WVTR of 1.14×10^{-5} g/m^2/day and an average transmittance of 85.8% in the visible region for the flexible multi-barrier containing a silica nanoparticle-embedded organic–inorganic hybrid (SCH) nanocomposite and Al_2O_3[93] and [94]. Moreover, they also studied theMgO/SCH hybrid structure. The results of the Ca corrosion test showed that 4.5 dyads of the MgO/SCH nanocomposite had extremely low WVTR of 4.33×10^{-6} g/m^2/day and an excellent optical transmittance of 84% [95]. Based on the above studies, in 2014, Duan et al. optimized the ALD/MLD nanolaminates by substituting H_2O by O_3 as the oxidant. Estimation of WVTR yielded significantly better results for the O_3-based laminates, with values decreasing linearly from 3.22×10^{-3} g/m^2/day to 2.37×10^{-5} g/m^2/day as the number of laminate layers increased from one to three, whereas a gentle decreasing trend from 1.83×10^{-3} g/m^2/day to 5.92×10^{-4} g/m^2/day was obtained for the H_2O-based laminate [77]. Besides, their adjustable composition allows the tuning of the optical properties, thereby enhancing their application potential for the design and fabrication of high-performance light out-coupling structures for top emitting OLEDs. By carefully adjusting the relative thickness ratio of the inorganic–organic encapsulation materials, optimized light extraction and high moisture barrier performance were achieved [79].

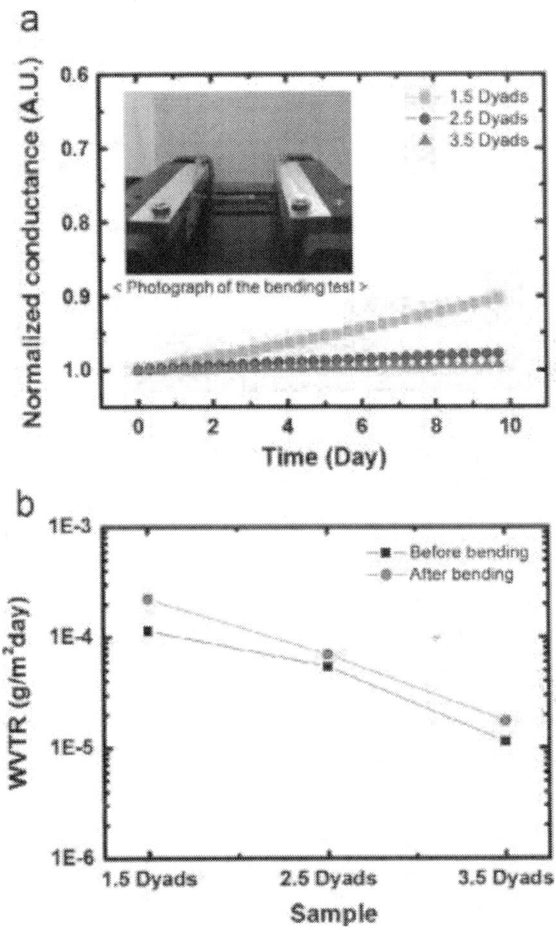

Figure 6. Ca test results after the bending test while varying the multi-barrier stacks; (a) normalized conductance versus time curve after 100 iterations of bending and (b) comparison of the WVTR values after the bending test. The inset shows the photograph of the bending test set-up.

Even after overcoming the water vapor permeation problem, another concern about FOLEDs exits. When a FOLED device suffers from an external stress, cracks or even delamination may appear among the thin films when the stress exceeds more than an optimum value. In a conventional OLED device structure, ITO is known as the most common anode material with Young's Modulus and yield stress of 120 GPa and 1.2 GPa, respectively, making it the most brittle layer in the device. Researchers have studied the mechanical properties of ITO under different types of stresses [96]. Tensile and compressive stress can make ITO thin film cracking or buckling. On account of this, the sheet resistance will increase, and the films will delaminate finally [97] and [98]. Although many researchers are working on finding a substitute for ITO, this material is still the best as the anode in OLED devices. Therefore, an encapsulation layer between the polymer substrate and ITO electrode with both efficient water vapor barrier performance and stress buffering effect is of great concern now [99].

SUMMARY

Organic electronic devices are the promising technologies in the future, especially for the flexible devices. To improve the stability, more advancing technologies are needed to encapsulate the devices instead of the traditional methods. Previous studies have shown the potential of the encapsulation property of the TFE technology. After depositing thin film by PECVD and ALD, the operating time of organic devices increased significantly. By optimizing the materials, structures of barriers and the deposition processes, the value of WVTR decreased significantly. Recently, the ALD/MLD technologies offer a great potential for the applications in flexible devices encapsulation. However, there are still issues in manufacturing, practically the TFE on large area/flexible application. This article discusses the present issue and the potential technologies to improve the barrier properties for organic electronics. The flexible devices with the development of TFE methods will be the promising technology in the future.

ACKNOWLEDGEMENTS

This study was supported by Program of International Science and Technology Cooperation (2014DFG12390), National High Technology Research and Development Program of China (Grant no. 2011AA03A110), Ministry of Science and Technology of China (Grant nos. 2010CB327701, 2013CB834802), National Natural Science Foundation of China(Grant nos. 61275024, 61377026, 61274002, 61275033, and 61177025), Scientific and Technological Developing Scheme of Jilin Province (Grant nos. 20140101204JC,

20130206020GX, 20140520071JH, 20130102009JC), Scientific and Technological Developing Scheme of Changchun (Grant no. 13GH02), Opened Fund of the State Key Laboratory on Integrated Optoelectronics (No. IOSKL2012KF01).

REFERENCES

1. M. Pope, H.P. Kallmann, P. Magnante, J. Chem. Phys. 38 (1963) 2042.
2. S. Coe, W.K. Woo, M. Bawendi, V. Bulovic, Nature 420 (2002) 800.
3. H. Shirakawa, E.J. Louis, A.G. MacDiarmid, C.K. Chiang, A.J. Heeger, J. Chem. Soc. Chem. Commun. (1977) 578.
4. C.W. Tang, Appl. Phys. Lett. 48 (1986) 183.
5. H.J. Bolink, H. Brine, E. Coronado, M. Sessolo, ACS Appl. Mater. Interfaces 2 (2010) 2694.
6. A. Tsumura, H. Koezuka, T. Ando, Appl. Phys. Lett. 49 (1986) 1210.
7. J.S. Park, H. Chae, H.K. Chung, S.I. Lee, Semicond. Sci. Technol. 26 (2011) 0340001.
8. Paul E. Burrows, Gordon L. Graff, Mark E. Gross, Peter M. Martin, M. Hall, Eric Mast, Charles C. Bonham, Wendy D. Bennett, Lech A. Michalski, Michael S. Weaver, Julie J. Brown, D. Fogarty, Linda S. Sapochak, Gas permeation and lifetime tests on polymer-based barrier coatings, Proc. SPIE. 4105 (2001) 75–83, http://dx.doi.org/10.1117/12.416878.
9. A.A. Dameron, S.D. Davidson, B.B. Burton, P.F. Carcia, R.S. McLean, S.M. George, J. Phys. Chem. C 112 (2008) 4573.
10. H. Aziz, Z.D. Popovic, N.X. Hu, A.M. Hor, G. Xu, Science 283 (1999) 1900.
11. J. Shen, D. Wang, E. Langlois, W.A. Barrow, P.J. Green, C.W. Tang, J. Shi, Synth. Met. 111 (2000) 233.
12. S.T. Lee, Z.Q. Gao, L.S. Hung, Appl. Phys. Lett. 75 (1999) 1404.
13. S.F. Lim, W. Wang, S.J. Chua, Mater. Sci. Eng. B—Solid State Mater. Adv. Technol. 85 (2001) 154.
14. M. Schaer, F. Nuesch, D. Berner, W. Leo, L. Zuppiroli, Adv. Funct. Mater. 11 (2001) 116.
15. H. Yuji, A. Chihaya, T. Tetsuo, S. Shogo, Jpn. J. Appl. Phys. 31 (1992) 1812.
16. N. Grossiord, J.M. Kroon, R. Andriessen, P.W.M. Blom, Org. Electron. 13 (2012) 432.
17. K. Fukuda, T. Yokota, K. Kuribara, T. Sekitani, U. Zschieschang, H. Klauk, T. Someya, Appl. Phys. Lett. 96 (2010) 053302.
18. J.A. Hauch, P. Schilinsky, S.A. Choulis, S. Rajoelson, C.J. Brabec, Appl. Phys. Lett. 93 (2008) 103306.

19. P.F. Carcia, R.S. McLean, M.H. Reilly, M.D. Groner, S.M. George, Appl. Phys. Lett. 89 (2006) 031915.
20. F.L. Wong, M.K. Fung, S.L. Tao, S.L. Lai, W.M. Tsang, K.H. Kong, W.M. Choy, C. S. Lee, S.T. Lee, J. Appl. Phys. 104 (2008) 014509.
21. R. Paetzold, A. Winnacker, D. Henseler, V. Cesari, K. Heuser, Rev. Sci. Instr. 74 (2003) 5147.
22. A.R. Coulter, R.A. Deeken, G.M. Zentner, J. Membr. Sci. 65 (1992) 269.
23. M.D. Groner, S.M. George, R.S. McLean, P.F. Carcia, Appl. Phys. Lett. 88 (2006) 051907.
24. C.M. Hansen, Prog. Org. Coat. 42 (2001) 167.
25. S. Cros, R. de Bettignies, S. Berson, S. Bailly, P. Maisse, N. Lemaitre, S. Guillerez, Sol. Energy Mater. Sol. Cells 95 (2011) S65.
26. M.D. Kempe, Sol. Energy Mater. Sol. Cells 90 (2006) 2720.
27. J.D. Affinito, S. Eufinger, M.E. Gross, G.L. Graff, P.M. Martin, Thin Solid Films 308 (1997) 19.
28. H.Y. Li, Y.F. Liu, Y. Duan, Y.Q. Yang, Y.-N. Lu, Materials 8 (2015) 600.
29. Y.Q. Yang, Y. Duan, P. Chen, F.B. Sun, Y.H. Duan, X. Wang, D. Yang, J. Phys. Chem. C 117 (2013) 20308.
30. P.E. Burrows, V. Bulovic, S.R. Forrest, L.S. Sapochak, D.M. McCarty, M. E. Thompson, Appl. Phys. Lett. 65 (1994) 2922.
31. J.S. Lewis, M.S. Weaver, IEEE J. Sel. Top. Quant. Electron. 10 (2004) 45.
32. M.S. Weaver, L.A. Michalski, K. Rajan, M.A. Rothman, J.A. Silvernail, J.J. Brown, P.E. Burrows, G.L. Graff, M.E. Gross, P.M. Martin, M. Hall, E. Mast, C. Bonham, W. Bennett, M. Zumhoff, Appl. Phys. Lett. 81 (2002) 2929.
33. D. Yang, Y.Q. Yang, Y. Duan, P. Chen, C.L. Zang, Y. Xie, D.M. Liu, X. Wang, Y. H. Duan, F.B. Sun, Q. Gao, K.W. Xue, E.C.S. Solid, State Lett. 2 (2013) R31.
34. J.S. Park, T.W. Kim, D. Stryakhilev, J.S. Lee, S.G. An, Y.S. Pyo, D.B. Lee, Y.G. Mo, D. U. Jin, H.K. Chung, Appl. Phys. Lett. 95 (2009) 013503.
35. P.F. Baude, F.B. Mccormick, G.D. Vernstrom, Encapsulated organic electronic devices and method for making same, Google Patents, 2002.
36. T.B. Harvey, S.Q. Shi, F. So, Organic light emitting diodes, Google Patents, 1998.
37. T.B. Harvey, S.Q. Shi, F. So, Hermetic sealing of plastic substrate, Google Patents, 1997.
38. E.H.H. Jamieson, A.H. Windle, J. Mater. Sci. 18 (1983) 64.
39. S.M. Jeong, W.H. Koo, S.H. Choi, H.K. Balk, Solid-State Electron. 49 (2005) 838.
40. F. Duerinckx, J. Szlufcik, Sol. Energy Mater. Sol. Cells 72 (2002) 231.
41. K. Jung, J.Y. Bae, S.J. Park, S. Yoo, B.S. Bae, J. Mater. Chem. 21 (2011) 1977.

42. H. Jung, H. Jeon, H. Choi, G. Ham, S. Shin, H. Jeon, J. Appl. Phys. 115 (2014).
43. D.S. Wuu, T.N. Chen, C.C. Wu, C.C. Chiang, Y.P. Chen, R.H. Horng, F.S. Juang, Chem. Vap. Depos. 12 (2006) 220.
44. D.S. Wuu, W.C. Lo, C.C. Chiang, H.B. Lin, L.S. Chang, R.H. Horng, C.L. Huang, Y. J. Gao, Surf. Coat. Technol. 197 (2005) 253.
45. J. Perrin, J. Schmitt, C. Hollenstein, A. Howling, L. Sansonnens, Plasma Phys. Control. Fusion 42 (2000) B353.
46. A.P. Ghosh, L.J. Gerenser, C.M. Jarman, J.E. Fornalik, Appl. Phys. Lett. 86 (2005) 223503.
47. A.G. Aberle, Sol. Energy Mater. Sol. Cells 65 (2001) 239.
48. A.G. Erlat, R.J. Spontak, R.P. Clarke, T.C. Robinson, P.D. Haaland, Y. Tropsha, N. G. Harvey, E.A. Vogler, J. Phys. Chem. B 103 (1999) 6047.
49. H.K. Kim, S.W. Kim, D.G. Kim, J.W. Kang, M.S. Kim, W.J. Cho, Thin Solid Films 515 (2007) 4758.
50. H. Lin, L. Xu, X. Chen, X. Wang, M. Sheng, F. Stubhan, K.H. Merkel, J. Wilde, Thin Solid Films 333 (1998) 71.
51. W.D. Huang, X.H. Wang, M. Sheng, L.Q. Xu, F. Stubhan, L. Luo, T. Feng, X. Wang, F.M. Zhang, S.C. Zou, Mater. Sci. Eng. B—Solid State Mater. Adv. Technol. 98 (2003) 248.
52. T. Suntola, J. Antson, Method for producing compound thin films, Google Patents, 1977.
53. M. Ritala, M. Leskela, J.P. Dekker, C. Mutsaers, P.J. Soininen, J. Skarp, Chem. Vap. Depos. 5 (1999) 7.
54. S.M. George, A.W. Ott, J.W. Klaus, J. Phys. Chem. 100 (1996) 13121.
55. R.L. Puurunen, J. Appl. Phys. 97 (2005) 121301.
56. M. Leskelä, M. Ritala, Angew. Chem. Int. Ed. 42 (2003) 5548.
57. S.M. George, Chem. Rev. 110 (2010) 111.
58. M.D. Groner, F.H. Fabreguette, J.W. Elam, S.M. George, Chem. Mater. 16 (2004) 639.
59. S. Sarkar, J.H. Culp, J.T. Whyland, M. Garvan, V. Misra, Org. Electron. 11 (2010) 1896.
60. Y.Q. Yang, Y. Duan, J. Phys. Chem. C 118 (2014) 18783.
61. Y.Q. Yang, Y. Duan, Y.H. Duan, X. Wang, P. Chen, D. Yang, F.B. Sun, K.w. Xue, Org. Electron. 15 (2014) 1120.
62. S.W. Seo, E. Jung, H. Chae, S.M. Cho, Org. Electron. 13 (2012) 2436.
63. S. Lee, H. Choi, S. Shin, J. Park, G. Ham, H. Jung, H. Jeon, Curr. Appl. Phys. 14 (2014) 552.

64. J. Meyer, P. Goerrn, F. Bertram, S. Hamwi, T. Winkler, H.-H. Johannes, T. Weimann, P. Hinze, T. Riedl, W. Kowalsky, Adv. Mater. 21 (2009) 1845.
65. L.H. Kim, K. Kim, S. Park, Y.J. Jeong, H. Kim, D.S. Chung, S.H. Kim, C.E. Park, ACS Appl. Mater. Interfaces 6 (2014) 6731.
66. A. Singh, F. Nehm, L. Mueller-Meskamp, C. Hossbach, M. Albert, U. Schroeder, K. Leo, T. Mikolajick, Org. Electron. 15 (2014) 2587.
67. J. Meyer, H. Schmidt, W. Kowalsky, T. Riedl, A. Kahn, Appl. Phys. Lett. 96 (2010) 243308.
68. B.H. Lee, V.R. Anderson, S.M. George, Atom. Layer Depos. Appl. 7 (41) (2011) 131.
69. P. Sundberg, M. Karppinen, Beilstein J. Nanotechnol. 5 (2014) 1104.
70. B.H. Lee, B. Yoon, V.R. Anderson, S.M. George, J. Phys. Chem. C 116 (2012) 3250.
71. B.H. Lee, V.R. Anderson, S.M. George, Chem. Vap. Depos. 19 (2013) 204.
72. M. Park, S. Oh, H. Kim, D. Jung, D. Choi, J.S. Park, Thin Solid Films 546 (2013) 153.
73. A. Sood, P. Sundberg, M. Karppinen, Dalton Trans. 42 (2013) 3869.
74. B.H. Lee, B. Yoon, A.I. Abdulagatov, R.A. Hall, S.M. George, Adv. Funct. Mater. 23 (2013) 532.
75. S.H. Jen, S.M. George, ACS Appl. Mater. Interfaces 5 (2013) 1165.
76. B.H. Lee, K.K. Im, K.H. Lee, S. Im, M.M. Sung, Thin Solid Films 517 (2009) 4056.
77. W. Xiao, D. Yu, S.F. Bo, Y.Y. Qiang, Y. Dan, C. Ping, D.Y. Hui, Z. Yi, RSC Adv. 4 (2014) 43850.
78. Y. Duan, X. Wang, Y.H. Duan, Y.Q. Yang, P. Chen, D. Yang, F.-B. Sun, K.W. Xue, N. Hu, J.W. Hou, Org. Electron. 15 (2014) 1936.
79. F.B. Sun, Y. Duan, Y.Q. Yang, P. Chen, Y.H. Duan, X. Wang, D. Yang, K.W. Xue, Org. Electron. 15 (2014) 2546.
80. S. Logothetidis, Mater. Sci. Eng. B—Adv. Funct. Solid-State Mater. 152 (2008) 96.
81. P. Schwamb, T. Reusch, C.J. Brabec, Organic Light Emitting Materials And Devices Xvii 8829, 2013.
82. Y. Leterrier, Prog. Mater. Sci. 48 (2003) 1–55.
83. A. Yoshida, A. Sugimoto, T. Miyadera, S. Miyaguchi, J. Photopolym. Sci. Technol. 14 (2001) 327.
84. H. Chatham, Surf. Coat. Technol. 78 (1996) 1–9.
85. C. Charton, N. Schiller, M. Fahland, A. Hollander, A. Wedel, K. Noller, Thin Solid Films 502 (2006) 99.
86. J.D. Affinito, M.E. Gross, C.A. Coronado, G.L. Graff, I.N. Greenwell, P.M. Martin, Thin Solid Films 290 (1996) 63.

87. J. Greener, K.C. Ng, K.M. Vaeth, T.M. Smith, J. Appl. Polym. Sci. 106 (2007) 3534.
88. A.B. Chwang, M.A. Rothman, S.Y. Mao, R.H. Hewitt, M.S. Weaver, J.A. Silvernail, K. Rajan, M. Hack, J.J. Brown, X. Chu, L. Moro, T. Krajewski, N. Rutherford, Appl. Phys. Lett. 83 (2003) 413.
89. J.W. Han, H.J. Kang, J.Y. Kim, G.Y. Kim, D.S. Seo, Jpn. J. Appl. Phys. Part 1—Regul. Pap. Brief Commun. Rev. Pap. 45 (2006) 9203.
90. D.S. Wuu, W.C. Lo, L.S. Chang, R.H. Horng, Thin Solid Films 468 (2004) 105.
91. J.W. Han, H.J. Kang, J.H. Kim, D.S. Seo, Jpn. J. Appl. Phys. Part 2—Lett. Express Lett. 45 (2006) L827.
92. M. Vaha-Nissi, P. Sundberg, E. Kauppi, T. Hirvikorpi, J. Sievanen, A. Sood, M. Karppinen, A. Harlin, Thin Solid Films 520 (2012) 6780.
93. Y.C. Han, E. Kim, W. Kim, H.G. Im, B.S. Bae, K.C. Choi, Org. Electron. 14 (2013) 1435.
94. S.W. Seo, H. Chae, S.J. Seo, H.K. Chung, S.M. Cho, Appl. Phys. Lett. 102 (2013) 161908.
95. E. Kim, Y. Han, W. Kim, K.C. Choi, H.G. Im, B.S. Bae, Org. Electron. 14 (2013) 1737.
96. P.C.P. Bouten, P.J. Slikkerveer, Y. Leterrier, Mechanics of ITO on Plastic Substrates for Flexible Displays, Flexible Flat Panel Displays, John Wiley & Sons, Ltd (2005), p. 99–120.
97. Y. Leterrier, L. Medico, F. Demarco, J.A.E. Manson, U. Betz, M.F. Escola, M. K. Olsson, F. Atamny, Thin Solid Films 460 (2004) 156.
98. C.J. Chiang, C. Winscom, A. Monkman, Org. Electron. 11 (2010) 1870.
99. C.J. Chiang, C. Winscom, S. Bull, A. Monkman, Org. Electron. 10 (2009) 1268.

CITATION

Duan Yu, Yong-Qiang Yang, Zheng Chen, Ye Tao, Yun-Fei Liu, Recent progress on thin-film encapsulation technologies for organic electronic devices, Optics Communications, Available online 25 August 2015, ISSN 0030-4018, http://dx.doi.org/10.1016/j.optcom.2015.08.021.

CHAPTER 2

A Quantum Energy Transport Model for Semiconductor Device Simulation

Shohiro Sho[1], *Shinji Odanaka*[2]

[1] Graduate School of Information Science and Technology, Osaka University, Osaka, Japan
[2] Computer Assisted Science Division, Cybermedia Center, Osaka University, Osaka, Japan

ABSTRACT

This paper describes numerical methods for a quantum energy transport (QET) model in semiconductors, which is derived by using a diffusion scaling in the quantum hydrodynamic (QHD) model. We newly drive a four-moments QET model similar with a classical ET model. Space discretization is performed by a new set of unknown variables. Numerical stability and convergence are obtained by developing numerical schemes and an iterative solution method with a relaxation method. Numerical simulations of electron transport in a scaled MOSFET device are discussed. The QET model allows simulations of quantum confinement transport, and nonlocal and hot-carrier effects in scaled MOSFETs.

INTRODUCTION

The semiconductor devices are scaled down into the nanoscale regime to achieve high circuit performance in the future integrated system. The performance of nanoscale semiconductor devices primarily relies on carrier transport properties in the short channels. Quantum energy transport (QET) models have been developed to understand such physical phenomena in scaled semiconductor devices. A full QET model has been derived from the collisional Wigner–Boltzmann equations using the entropy minimization principle [1]. Numerical simulations using this model, however, have not been performed [2]. Simplified QET models have been proposed as the energy transport extension of the quantum drift diffusion (QDD) model with Fourier law closure and numerically investigated [3] and [4]. In Ref. [4], the carrier temperature in the current density is further approximated by the lattice temperature to bring the model into a self-adjoint form.

In this paper, we develop numerical methods for a QET model derived from a quantum hydrodynamic (QHD) model. To overcome the difficulties associated with the Fourier law closure, we newly derive a four-moments QET model similar with a classical energy transport (ET) model [5]. The numerical stability is achieved by developing numerical schemes and an iterative solution method in terms of a new set of variables. Numerical results in a scaled MOSFET are demonstrated.

The paper is organized as follows: In Section 2, a four-moments QET model is derived from the QHD model. In Section 3, we present nonlinear discretization schemes and an iterative solution method to solve the QET system. In Section 4, numerical simulations of electron transport in a scaled MOSFET are discussed. Some conclusions are addressed in Section 5.

4 MOMENTS QUANTUM ENERGY TRANSPORT MODEL

The QET models are obtained by using a diffusion scaling in the quantum hydrodynamic equations, similar as in the classical hydrodynamic model [5]. The QHD model has been derived from the collisional Wigner-Boltzmann equations, assuming Fourier law closure [6]. For classical hydrodynamic simulations, the closure relation based on the four-moments of the Boltzmann equation has been discussed [7], [8] and [9], and the four-moments ET models are developed for simulations of thin body MOSFETs [5] and [10]. In this work, we derive a four-moments QET model from four moments equations derived from the collisional Wigner–Boltzmann equation.

For simplicity, we consider only the case of electrons. The four moment equations have the same form as the classical hydrodynamic equations [7],

$$\partial_t n+\nabla\cdot(n\mathbf{v})=nC_n, \tag{1}$$

$$\partial_t(n\mathbf{p})+\nabla\cdot(nU)-n\mathbf{F}_E=nC_p, \tag{2}$$

$$\partial_t(nw)+\nabla\cdot(n\mathbf{S})-n\mathbf{v}\cdot\mathbf{F}_E=nC_\epsilon, \tag{3}$$

$$\nabla\cdot(nR)-n(wI+U)\cdot\mathbf{F}_E=nC_{p\epsilon}, \tag{4}$$

where n,**p**, and w are the electron density, momentum, and kinetic energy, respectively. **v**,U,**S** and R are the velocity, second moment tensor, energy flow, and fourth moment tensor, respectively. I is the identity tensor. $F_E=-qE$, where E is the electric field. C_n,C_p,C_ϵ, and $C_{p\epsilon}$ are the electron generation rate, the production of crystal momentum, the energy production, and the production of the energy flux, respectively. (1) and (2), (3) and (4) represent conservation of particles, momentum, energy, and energy flux, respectively. By assuming parabolic bands, we give the following closure relations for **p** and U as

$$\mathbf{p}=m\mathbf{v}, \tag{5}$$

$$U_{ij}=mv_iv_j-\frac{P_{ij}}{n}, \tag{6}$$

where m is an effective mass. The quantum correction to the stress tensor P_{ij} was proposed by Ancona and Iafrate [11], and the quantum correction to the energy density$W=nw$ was first derived by Wigner [12], which are given by

$$P_{ij}=-nkT_n\delta_{ij}+\frac{\hbar^2}{12m}n\frac{\partial^2}{\partial x_i\partial x_j}\log n+O(\hbar^4), \tag{7}$$

$$W=\frac{1}{2}mnv^2+\frac{3}{2}nkT_n-\frac{\hbar^2}{24m}n\frac{\partial^2}{\partial x_k^2}\log n+O(\hbar^4), \tag{8}$$

where T_n and $\hbar$ are the electron temperature and Plank's constant, respectively.

For the collision terms, we employ a macroscopic relaxation time approximation to drive a QET model as follows:

$$C_n=0, \tag{9}$$

$$C_p=-\frac{p}{\tau_p}, \tag{10}$$

$$C_\epsilon=-\frac{w-w_0}{\tau_\epsilon}, \tag{11}$$

where τ_p and τ_ϵ are the momentum and energy relaxation times, respectively. Substituting (5), (6) and (7) into (1) and (2), we obtain moment equations for conservation of electron number and momentum

$$\frac{\partial n}{\partial t}+\frac{\partial}{\partial x_i}(nv_i)=0, \tag{12}$$

$$\frac{\partial}{\partial t}(mnv_i)+\frac{\partial}{\partial x_j}\left(mnv_iv_j+knT_n-\frac{\hbar^2}{12m}n\frac{\partial^2}{\partial x_i\partial x_j}\log n\right)=-n\frac{\partial V}{\partial x_i}-\frac{mnv_i}{\tau_p}. \tag{13}$$

We further get the following relation:

$$\frac{\partial}{\partial x_i} n \frac{\partial^2}{\partial x_i \partial x_j} \log n = 2n \frac{\partial}{\partial x_j} \frac{1}{\sqrt{n}} \frac{\partial^2}{\partial x_i^2} \sqrt{n}. \tag{14}$$

With the relation (14), the quantum correction term in (13) is written as

$$-\frac{\hbar^2}{12m} \frac{\partial}{\partial x_i} n \frac{\partial^2}{\partial x_i \partial x_j} \log n = -\frac{\hbar^2 n}{6m} \frac{\partial}{\partial x_i} \left(\frac{1}{\sqrt{n}} \frac{\partial^2}{\partial x_j^2} \sqrt{n} \right) = -qn \frac{\partial}{\partial x_i} \gamma_n, \tag{15}$$

where the term

$$\gamma_n = \frac{\hbar^2}{6mq} \frac{1}{\sqrt{n}} \frac{\partial^2}{\partial x_j^2} \sqrt{n} \tag{16}$$

is the quantum potential. Then, the conservation of momentum is given by

$$\frac{\partial}{\partial t}(mn v_i) + \frac{\partial}{\partial x_j}(mn v_i v_j + knT_n) - qn \frac{\partial}{\partial x_i} \gamma_n = -n \frac{\partial V}{\partial x_i} - \frac{mn v_i}{\tau_p}. \tag{17}$$

We can define the current density and the electric charge as $J_j = -qnv_j$ and q is the positive electric charge. Using a diffusion scaling in (17), we obtain

$$\tau_p \frac{\partial}{\partial t} J_i - k\mu_n \frac{\partial}{\partial x_i}(nT_n) + qn\mu_n \frac{\partial}{\partial x_i} \gamma_n = \mu_n n \frac{\partial V}{\partial x_i} - J_i, \tag{18}$$

where $\mu_n = +\frac{q\tau_p}{m}$ is the electron mobility. The potential energy is given by

$$V=-q\varphi. \tag{19}$$

From (12) and (18) and (19), we obtain the current continuity equation as follows:

$$\frac{1}{q}\operatorname{div} J_n = 0, \tag{20}$$

$$J_n = q\mu_n\left(\nabla\left(n\frac{kT_n}{q}\right) - n\nabla(\varphi + \gamma_n)\right). \tag{21}$$

The energy balance equation is derived from (3) and (4) [7]. The collision term in (2) is rewritten as

$$C_p = -\frac{qv}{\mu_n}. \tag{22}$$

In analogy to (22), the collision term in (4) is modeled as

$$C_{p\epsilon} = -\frac{qS}{\mu_s}, \tag{23}$$

where μ_s is the energy flow mobility. Neglecting the time derivative term in (2), we get

$$n\mathbf{F}_E = \nabla \cdot (nU) + n\frac{q\mathbf{V}}{\mu_n}. \tag{24}$$

Substituting (24) into (4), the expression of energy flux S is given as

$$S = \frac{\mu_s}{\mu_n}(wI + U)\cdot\mathbf{v} + \frac{\mu_s}{qn}((wI + U)\cdot\nabla\cdot(nU) - \nabla\cdot(nR)). \tag{25}$$

Assuming a heated Maxwellian distribution, the fourth moment tensor R is specified by the classical form as

$$R = \frac{5}{2}k^2 T_n^2 I. \tag{26}$$

Using closure (26), an expression for the energy flux density $S_n = nS$ is obtained as

$$S_n = \frac{\mu_s}{\mu_n}(WI + nU)\cdot\mathbf{v} + \frac{\mu_s}{q}\left((wI + U)\cdot\nabla\cdot(nU) - \nabla\cdot\left(\frac{5}{2}nk^2T_n^2 I\right)\right). \tag{27}$$

The second term of (27) is the diffusive contributions to the energy flux density which includes the classical form of R. In this work, we develop a QET model, neglecting quantum corrections in the diffusive contributions to the energy flux density. Substituting (6), (7) and (8) into (27), the quantum corrections to the energy density W and stress tensor P_{ij} are included in the drift contributions to the energy flux density S_n and neglected in the diffusive contributions. As a result, we obtain a quantum corrected expression for the energy flux density as

$$S_n = -\frac{\mu_s}{\mu_n}\left(\frac{5}{2}\frac{kT_n}{q} - \frac{\hbar^2}{24mq}\Delta\log n - \frac{\hbar^2}{12mq}\frac{\partial^2}{\partial x_i \partial x_j}\log n\right)J_n - \frac{\mu_s}{\mu_n}\frac{5}{2}\left(\frac{k}{q}\right)^2 q\mu_n n T_n \nabla T_n. \tag{28}$$

From (3), we get

$$\nabla\cdot S_n = -J_n\cdot\nabla\varphi - \frac{3}{2}kn\frac{T_n - T_L}{\tau_\varepsilon}. \tag{29}$$

Assuming that the velocity v is slowly varying in the device region, the following term in(29) is approximated as

$$\frac{\hbar^2}{12m}\frac{\partial}{\partial x_i}\left(nv_j\frac{\partial^2}{\partial x_i \partial x_j}\log n\right)=\frac{\hbar^2}{12m}v_j\frac{\partial}{\partial x_i}\left(n\frac{\partial^2}{\partial x_i \partial x_j}\log n\right)+\frac{\hbar^2}{12m}\frac{\partial v_j}{\partial x_i}\left(n\frac{\partial^2}{\partial x_i \partial x_j}\log n\right)\approx -J_n\frac{\partial}{\partial x_j}\gamma_n. \quad (30)$$

Then, we obtain a four-moments QET model as follows:

$$\epsilon\Delta\varphi=q(n-p-C), \quad (31)$$

$$\frac{1}{q}div J_n = 0, \quad (32)$$

$$J_n = q\mu_n\left(\nabla\left(n\frac{kT_n}{q}\right)-n\nabla(\varphi+\gamma_n)\right), \quad (33)$$

$$b_n\nabla\cdot(\rho_n\nabla u_n)-\frac{kT_n}{q}\rho_n u_n = -\frac{\rho_n}{2}(\varphi-\varphi_n), \quad (34)$$

$$\nabla\cdot S_n = -J_n\cdot\nabla\varphi-\frac{3}{2}kn\frac{T_n-T_L}{\tau_\epsilon}, \quad (35)$$

$$S_n = -\frac{\mu_s}{\mu_n}\left(\frac{5}{2}\frac{kT_n}{q}-\frac{\hbar^2}{24mq}\Delta\log n-\gamma_n\right)J_n-\frac{\mu_s}{\mu_n}\frac{5}{2}\left(\frac{k}{q}\right)^2 q\mu_n nT_n\nabla T_n, \quad (36)$$

where $v_n=\frac{(\varphi+\gamma_n-\varphi_n)}{2}$ and $u_n=\frac{q}{kT_n}v_n$. φ,φ_n, and p are the electrostatic potential, chemical potential, and hole density, respectively. ρ_n is the the root-density of electrons. ϵ,q, and k are the permittivity of semiconductor, electronic charge, and Boltzmann's constant. C and T_L are the ionized impurity density and the lattice temperature, respectively. The value of effective mass is given by a single parameterm=0.26m_0 in the silicon devices, where m_0 is the mass of a stationary electron. The quantum parameter for electrons becomes

$$b_n = \frac{\hbar^2}{12qm}. \tag{37}$$

For a temperature dependent mobility model, we apply the simplified Hänsch's mobility model [5],

$$\frac{\mu(T_n)}{\mu_0} = \left(1 + \frac{3}{2}\frac{\mu_0 k}{q\tau_\epsilon v_s^2}(T_n - T_L)\right)^{-1}, \tag{38}$$

where μ_0 and v_s are the low-field mobility and saturation velocity, respectively.

From (16), the quantum potential equation is derived as

$$2b_n\nabla^2\rho_n - \gamma_n\rho_n = 0. \tag{39}$$

In our model, (39) is replaced by (34) with respect to the variable u_n by employing an exponential transformation of variable $\rho_n = \sqrt{n} = \sqrt{n_i}exp(\frac{q}{kT_n}v_n)$[13]. If the variable $u_n$is uniformly bounded, the electron density is maintained to be positive. As mentioned below, this approach provides a numerical advantage for developing the iterative solution method of the QET model as well as the QDD model [13].

The system (31), (32), (33), (34), (35) and (36) are solved in the bounded domain Ω. The boundary $\partial\Omega$ of the domain Ω splits into two disjoint part Γ_D and Γ_N. The contacts of semiconductor devices are modeled by the boundary conditions on Γ_D, which fulfill charge neutrality and thermal equilibrium. We further assume that no quantum effects occur at the contacts. Here, the boundary conditions are given as follows:

$$\varphi = \varphi_b + \varphi_{appl}, \quad n = n_D, \quad u_n = u_D, \quad T_n = T_L \text{ on } \Gamma_D, \tag{40}$$

$$\nabla\varphi\cdot\nu = \nabla J_n\cdot\nu = \nabla u_n\cdot\nu = \nabla S_n\cdot\nu = 0 \text{ on } \Gamma_N, \tag{41}$$

where φ_b is a built-in potential and $\varphi appl$ is an applied bias voltage. $u_D = \frac{q}{kT_L}\frac{\varphi_b}{2}$ on the contacts and $u_n = u_0$, where u_0 is a small positive constant at the silicon dioxide interface.

DISCRETIZATION AND ITERATIVE SOLUTION METHOD

Discretization

Space discretization of the four-moments QET model is performed by a new set of unknown variables (φ, u_n, n, T_n). For the current density, we have

$$J_n = q\mu_n\left(\nabla\left(n\frac{kT_n}{q}\right) - \frac{q}{kT_n}\left(n\frac{kT_n}{q}\right)\nabla(\varphi + \gamma_n)\right). \tag{42}$$

As pointed out in discretization of classical hydrodynamic models [17] and [18], the total energy flow $H = S_n + \varphi J_n$, which consists of both the thermal energy flow S_n and the electrical flow φJ_n, is used to solve the energy balance equation. The total energy flow can be rewritten as

$$H = S_n + \varphi J_n = \widetilde{S}_n + \left(\varphi + \frac{\mu_s}{\mu_n}\left(\frac{\hbar^2}{24mq}\Delta \log n + \gamma_n\right)\right)J_n, \tag{43}$$

$$\widetilde{S}_n = -\frac{5}{2}\frac{\mu_s}{\mu_n}\frac{kT_n}{q}J_n - \frac{5}{2}\frac{\mu_s}{\mu_n}\left(\frac{k}{q}\right)^2 q\mu_n n T_n \nabla T_n. \tag{44}$$

Substituting (33) into (44), for the energy flow, we have

$$\begin{aligned}\widetilde{S}_n &= -\frac{5}{2}\frac{\mu_s}{\mu_n}q\mu_n\left(\frac{kT_n}{q}\nabla n\frac{kT_n}{q} - \frac{kT_n}{q}n\nabla(\varphi + \gamma_n) + \frac{kT_n}{q}n\nabla\frac{kT_n}{q}\right)\\ &= -\frac{5}{2}q\mu_s\left(\nabla n\left(\frac{kT_n}{q}\right)^2 - \frac{q}{kT_n}n\left(\frac{kT_n}{q}\right)^2\nabla(\varphi + \gamma_n)\right).\end{aligned} \tag{45}$$

When the variable ξ is defined as $\xi = n\frac{kT_n}{q} = n\eta$ in the current density J_n and $\xi = n(\frac{kT_n}{q})^2 = n\eta^2$ in the energy flow $\tilde{S}_n$, J_n and $\tilde{S}_n$ can be written in the same form, similar as in the classical ET models [10] and [14],

$$\nabla \cdot F = \nabla \cdot \left(C \left(\nabla\xi - \frac{q}{kT_n}\xi\nabla(\varphi + \gamma_n) \right) \right), \tag{46}$$

where F is the flux. The constant C is defined as $C=q\mu_n$ in J_n and $C = -\frac{5}{2}q\mu_s$ in $\tilde{S}_n$. By projecting (46) onto a grid line and using the variable $g = \int_{x_i}^{x} \frac{q}{kT_n}\nabla(\varphi + \gamma_n)$, a one-dimensional self-adjoint form is obtained as

$$\frac{d}{dx}F = \frac{d}{dx}(Ce^{g}\frac{d}{dx}(e^{-g}\xi)). \tag{47}$$

For space discretization, the simulation region is divided into computational cells Ωijcentered at (x_i,y_j). In a staggered Cartesian grid, each computational cell is rectangular, and the variables φ, u_n, n, T_n are defined at cell centers and the flux is defined at cell interfaces. For space discretization of (47), we construct high-accuracy nonlinear schemes, applying the finite-volume method to construct multidimensional schemes. For the flux $F=Ce^{g}\nabla(e^{-g}\xi)$, we integrate (47) over the computational cellsΩij. Using Green's theorem, we obtain a discrete form as

$$\int_{\Omega_{ij}} \nabla \cdot F dx = a_j\left(F_{i+\frac{1}{2}} - F_{i-\frac{1}{2}}\right) + a_i\left(F_{j+\frac{1}{2}} - F_{j-\frac{1}{2}}\right), \tag{48}$$

where a_i and a_j are the cell sizes of the computational cell Ωij. In order to find $F_{i+\frac{1}{2}}$ at cell interfaces, integrating the flux F over the interval $[x_i,x_{i+1}]$, an approximation $F_{i+\frac{1}{2}}$ yields

$$F_{i+\frac{1}{2}} = \frac{C(\psi_{i+1,j} - \psi_{i,j})}{\int_{x_i}^{x_{i+1}} e^{-g}dx}, \tag{49}$$

where $\psi=e^{-g}\xi$. A similar expression is obtained for $F_{i-\frac{1}{2}}, F_{j+\frac{1}{2}}$, and $F_{j-\frac{1}{2}}$. The accuracy of the numerical flux depends on the explicit integration $\int_{x_i}^{x_{i+1}} e^{-g} dx$ in (49). In order to construct a higher accuracy nonlinear scheme, an explicit integration $\int_{x_i}^{x_{i+1}} e^{-g} dx$ is obtained by the piecewise linear approximation of φ and T_n on the interval $[x_i,x_{i+1}]$[15] and [16]. Then we have

$$F_{i+\frac{1}{2}} = \frac{C}{\theta_{i+1}^x h_{i+1}^x}\left(B(\Delta_{i+1}^x)\frac{\xi_{i+1,j}^x}{\eta_{i+1,j}} - B(-\Delta_{i+1}^x)\frac{\xi_{i,j}^x}{\eta_{i,j}}\right), \tag{50}$$

where B(·) is the Bernoulli function. h_{i+1}^x is defined as $h_{i+1}^x = (a_{i+1}^x + a_i^x)/2$. The variables $\theta_{i+1}^x, \Delta_{i+1}^x$ are calculated as follows:

$$\theta_{i+1}^x = \log\left(\frac{\eta_{i+1,j}}{\eta_{i,j}}\right)/(\eta_{i+1,j} - \eta_{i,j}), \tag{51}$$

$$\Delta_{i+1}^x = \theta_{i+1}^x((\varphi_{i+1,j} - \varphi_{i,j}) + (\gamma_{n_{i+1,j}} - \gamma_{n_{i,j}}) - (\eta_{i+1,j} - \eta_{i,j})). \tag{52}$$

Such schemes to J_n and $\widetilde{S}_n$ result in a consistent generalization of the Scharfetter–Gummel type schemes to the QET equations. The energy balance equation is further discretized using (49). To conserve the total energy flow $H=S_n+\varphi J_n$ (43), discretization of the carrier heating term is another key issue [17] and [18]. Integrating(35) over the computational cell yields

$$\int_{\Omega_{ij}} \nabla\cdot\widetilde{S}_n dx$$
$$= \int_{\Omega_{ij}} -J_n\cdot\nabla\left(\varphi + \frac{\mu_s}{\mu_n}\left(\gamma_n + \frac{b_n}{2}\Delta\log n\right)\right)dx - \int_{\Omega_{ij}} \frac{3}{2}kn\frac{T_n - T_L}{\tau_\epsilon}dx. \tag{53}$$

Here, quantum corrections are included in the carrier heating term. From Gauss's theorem, the first term on the right hand side of (53) can be calculated as

$$\int_{\Omega_{ij}} -J_n\cdot\nabla\left(\varphi + \frac{\mu_s}{\mu_n}\left(\gamma_n + \frac{b_n}{2}\Delta\log n\right)\right)dx = -\int_{\partial\Omega_{ij}}\left(J_n\left(\varphi + \frac{\mu_s}{\mu_n}\left(\gamma_n + \frac{b_n}{2}\Delta\log n\right)\right)\right)\cdot\nu dx. \tag{54}$$

Assuming the Boltzmann statics, the electron density is expressed as

$$n = n_i \exp\left(\frac{q(\varphi + \gamma - \varphi_n)}{kT_n}\right) = n_i \exp(2u_n), \tag{55}$$

where n_i is the intrinsic density. Then, the discretization for $\Delta \log n = 2\Delta u_n$ in (54) is obtained by a standard five-point approximation:

$$\Delta^h u_n^h = \frac{1}{a_j^y h_{j+1}^y} u_{i,j+1} + \frac{1}{a_j^y h_j^y} u_{i,j-1} + \frac{1}{a_i^x h_{i+1}^x} u_{i+1,j} + \frac{1}{a_i^x h_i^x} u_{i-1,j} - \left(\frac{h_{j+1}^y + h_j^y}{a_j^y h_{j+1}^y h_j^y} + \frac{h_{i+1}^x + h_i^x}{a_i^x h_{i+1}^x h_i^x}\right) u_{i,j}. \tag{56}$$

The discrete form of the carrier heating term in (53) yields

$$\int_{\Omega_{ij}} -J_n \cdot \nabla\left(\varphi + \frac{\mu_s}{\mu_n}(\gamma_n + b_n \Delta^h u_n^h)\right) dx \approx -a_i^x\left(J_{n_{j+\frac{1}{2}}}\left(\varphi_{j+\frac{1}{2}} + \frac{\mu_s}{\mu_n}\left(\gamma_{n_{j+\frac{1}{2}}} + b_n \Delta^h u_n^h\right)\right)\right.$$
$$\left.-J_{n_{j-\frac{1}{2}}}\left(\varphi_{j-\frac{1}{2}} + \frac{\mu_s}{\mu_n}\left(\gamma_{n_{j-\frac{1}{2}}} + b_n \Delta^h u_n^h\right)\right)\right) - a_j^y\left(J_{n_{i+\frac{1}{2}}}\left(\varphi_{i+\frac{1}{2}} + \frac{\mu_s}{\mu_n}\left(\gamma_{n_{i+\frac{1}{2}}} + b_n \Delta^h u_n^h\right)\right) - J_{n_{i-\frac{1}{2}}}\left(\varphi_{i-\frac{1}{2}} + \frac{\mu_s}{\mu_n}\left(\gamma_{n_{i-\frac{1}{2}}} + b_n \Delta^h u_n^h\right)\right)\right). \tag{57}$$

Space discretization of (34) is performed following our previous works [13] and [19] to achieve a Scharfetter–Gummel type scheme, i.e.,

$$\frac{a_j^y}{h_{i+1}^x} b_n e^{u_{n_{i+1,j}}} B(u_{n_{i+1,j}} - u_{n_{i,j}})(u_{n_{i+1,j}} - u_{n_{i,j}}) - \frac{a_j^y}{h_i^x} b_n e^{u_{n_{i,j}}} B(u_{n_{i,j}} - u_{n_{i-1,j}})(u_{n_{i,j}} - u_{n_{i-1,j}}) + \frac{a_i^x}{h_{j+1}^y} b_n e^{u_{n_{i,j+1}}} B(u_{n_{i,j+1}} - u_{n_{i,j}})(u_{n_{i,j+1}}$$
$$- u_{n_{i,j}}) - \frac{a_j^x}{h_j^y} b_n e^{u_{n_{i,j}}} B(u_{n_{i,j}} - u_{n_{i,j-1}})(u_{n_{i,j}} - u_{n_{i,j-1}}) - \eta_{ij} u_{n_{ij}} \Lambda_{ij}$$
$$= -\frac{1}{2}(\varphi_{ij} - \varphi_{m_{ij}})\Lambda_{ij}, \tag{58}$$

where $\Lambda ij = \int_\Omega ij \rho_n dx$, which is approximated as

$$\Lambda_{ij} = \frac{1}{4} e^{u_{n_{ij}}} \times \left(\frac{h_i^x h_j^y}{B\left(\frac{u_{i-1,j} - u_{i,j}}{2}\right) B\left(\frac{u_{i,j-1} - u_{i,j}}{2}\right)} + \frac{h_{i+1}^x h_j^y}{B\left(\frac{u_{i+1,j} - u_{i,j}}{2}\right) B\left(\frac{u_{i,j-1} - u_{i,j}}{2}\right)} + \frac{h_i^x h_{j+1}^y}{B\left(\frac{u_{i-1,j} - u_{i,j}}{2}\right) B\left(\frac{u_{i,j+1} - u_{i,j}}{2}\right)} + \frac{h_{i+1}^x h_j^y}{B\left(\frac{u_{i+1,j} - u_{i,j}}{2}\right) B\left(\frac{u_{i,j+1} - u_{i,j}}{2}\right)}\right). \tag{59}$$

Iterative solution method

We develop an iterative solution method of the QET model by constructing a Gummel map [20] with a new set of unknown variables (φ, u_n, n, T_n) as follows:

(P1) Let $\varphi^m, n^m, p^m, T_n^m$ are given, solve the nonlinear Poisson equation with respect to the electrostatic potential φ^{m+1}, where m is the number of iteration. Eq. (31) is linearized using a Newton method. Then the linearized equation becomes

$$\epsilon\Delta\varphi^{m+1} - \frac{q^2}{k}\left(\frac{n^m}{T_n^m} + \frac{p}{T_p}\right)\varphi^{m+1} = q(n^m - p^m - C) - \frac{q^2}{k}\left(\frac{n^m}{T_n^m} + \frac{p}{T_p}\right)\varphi^m. \tag{60}$$

(P2) Let $\varphi^{m+1}, S^m, \varphi_n^m, T_n^m$ are given, solve the potential u_n^{m+1}.

$$b_n\nabla\cdot(\rho_n^m\nabla u_n^{m+1}) - \eta^m\rho_n^m u_n^{m+1} = -\frac{\rho_n^m}{2}(\varphi^{m+1} - \varphi_n^m). \tag{61}$$

Then, using u_n^{m+1} the quantum potential is further calculated as

$$\gamma_n^{m+1} = 2\eta^m u_n^{m+1} + \varphi_n^m - \varphi^{m+1}. \tag{62}$$

(P3) Let $\varphi^{m+1}, \gamma_n^{m+1}, T_n^m$ are given, solve the electron density n^{m+1}.

$$\frac{1}{q}div J_n = 0, \tag{63}$$

$$J_n = q\mu_n e^g\nabla(e^{-g}n^{m+1}\eta^m). \tag{64}$$

We set the generalized chemical potential by

$$\varphi_n^m = -\eta^m\log\frac{n^{m+1}}{n_i} + \varphi^{m+1} + \gamma_n^{m+1}. \tag{65}$$

(P4) Let $\varphi^{m+1}, \gamma_n^{m+1}, n^{m+1}, T_n^m$ are given, solve the electron temperature T_n^{m+1}.

$$\nabla \cdot \tilde{S}_n + \frac{3}{2}k\frac{n^{m+1}T_n^{m+1}}{\tau_\epsilon}$$
$$= -J_n \cdot \nabla\left(\varphi^{m+1} + \frac{\mu_s}{\mu_n}(\gamma_n^{m+1} + b_n \Delta u_n^{m+1})\right) + \frac{3}{2}k\frac{n^{m+1}T_L}{\tau_\epsilon}. \quad (66)$$

An iterative solution method, which consists of the inner and outer iteration loops, is developed, as shown in Fig. 1. The algorithm using the variable u_n in (34) ensures the positivity of the root-density of electrons without introducing damping parameters [13]. In fact, it is a critical issue to solve for the root-density ρ_n the quantum potential equation

$$-2b_n\nabla^2\rho_n+\gamma_n\rho_n=0. \quad (67)$$

In this case, the iterative solution method requires an additional iteration loop to maintain positive solutions for the root-density of electrons in the inner iteration loop as pointed out in Ref. [21]. Hence, in the inner iteration loop, (67) is replaced by (34). Therefore, we can enhance the robustness of the iterative solution method by introducing an under relaxation method with a parameter α, $0 < \alpha < 1$, in the outer iteration loop:

$$T^{m+1} = T^m + \alpha(T_*^{m+1} - T^m). \quad (68)$$

The convergence behavior of electron temperature is shown in Fig.2 as a function of the relaxation parameter. It is clear that the numerical stability is obtained by the relaxation method.

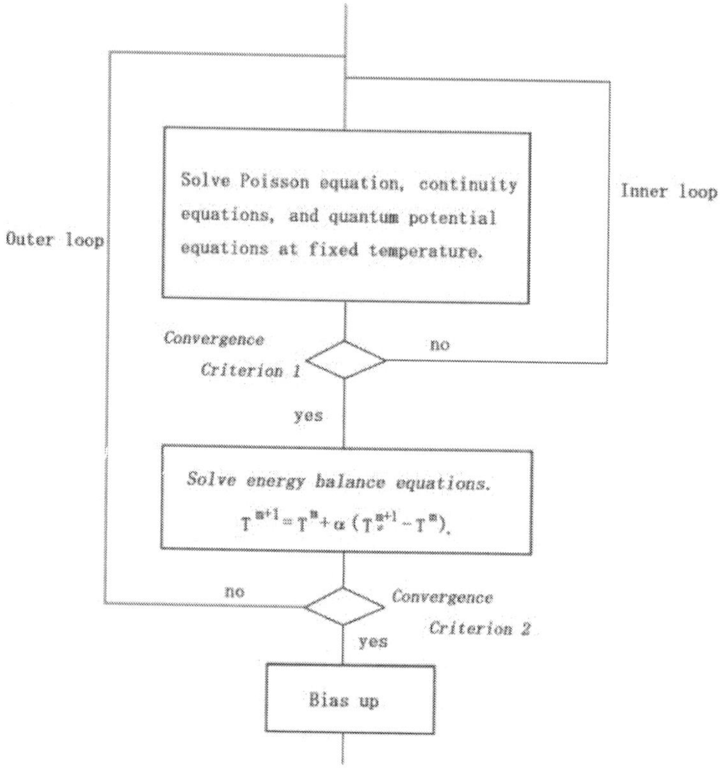

Figure 1. An iterative solution method with a relaxation algorithm.

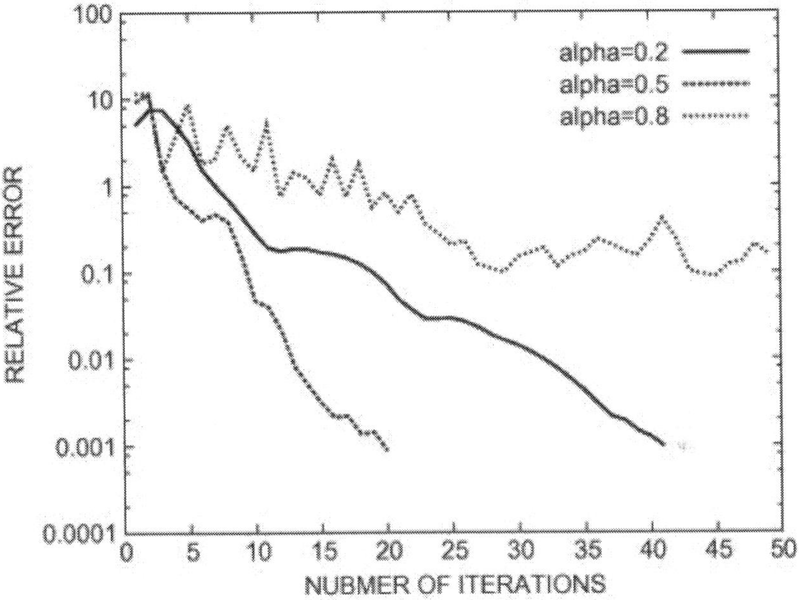

Figure 2. Relative error of electron temperature vs. number of iterations at different relaxation parameters.

NUMERICAL RESULTS

The numerical results are obtained for a 35 nm MOSFET having thin gate oxide thickness of 1.5 nm, uniform substrate concentration of 2.0×10^{18} cm^{-3}, and n-type doping concentration of 1.0×10^{20} cm^{-3}. The energy relaxation time τ_ϵ of $0.1{\times}10^{-12}$ ps and a ratio μ_s/μ_n of 0.8 are chosen. The MOSFET structure is shown in Fig. 3. The QET model includes a two-dimensional calculation of the electrostatic potential in the region with boundary A-G-L-F, and a two-dimensional calculation of the variables n,u_n, and $T_n$in the silicon region with boundary A-B-E-F. The mixed boundary conditions for the QET system are assigned as follows:

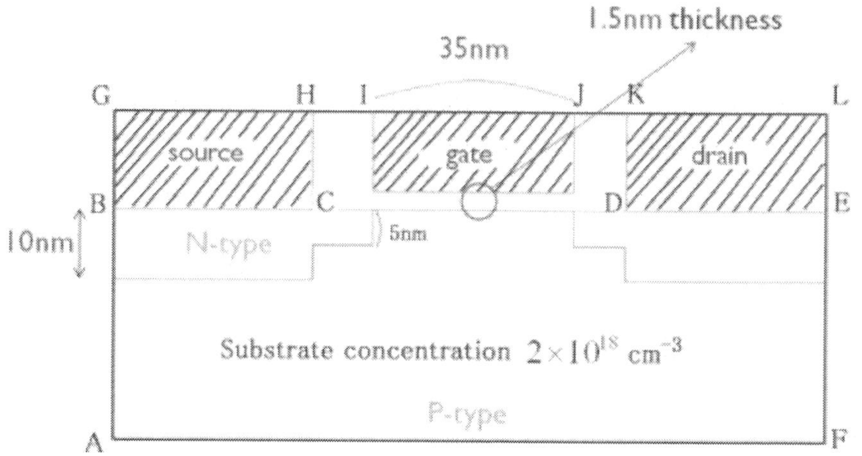

Figure 3. Two-dimensional cross section of a 35 nm MOSFET.

For the electrostatic potential φ

$$\varphi=\varphi appl+\varphi_b, \tag{69}$$

at source and drain regions, and back gate, where $\varphi appl$ is the applied bias voltage, and φ_b is the built-in potential, respectively. The gate region is also treated as a Dirichlet boundary condition with an approximated work function of the material. At the sides A–B, H–I, J–K, E–F, we have the homogeneous Neumann condition

$$\frac{\partial\varphi}{\partial v}=0. \tag{70}$$

For the variables n,u_n, and T_n, we have the constant Dirichlet conditions

$$n = \begin{cases} (C + \sqrt{C^2 + 4n_i^2})/2 \text{ at sides } B\text{–}C \text{ and } D\text{–}E, \\ 2n_i^2/(-C + \sqrt{C^2 + 4n_i^2}) \text{ at the back gate,} \end{cases}$$
$$T_n = T_L \text{ at sides } B\text{–}C,\ D\text{–}E, \text{ and } A\text{–}F,$$
$$u_n = \begin{cases} (q\varphi_b)/(2kT_n) \text{ at sides } B\text{–}C,\ D\text{–}E, \text{ and } A\text{–}F, \\ u_0 \text{ at the silicon-oxide interface } C\text{–}D, \end{cases} \quad (71)$$

where u_0 is the small positive constant. At the sides A–B and E–F, the homogeneous Neumann conditions read:

$$\frac{\partial n}{\partial v} = \frac{\partial T_n}{\partial v} = \frac{\partial u_n}{\partial v} = 0, \quad (72)$$

at the side C–D,

$$\frac{\partial n}{\partial v} = \frac{\partial T_n}{\partial v} = 0. \quad (73)$$

In Fig. 4 and Fig. 5, we compare the electron density distributions calculated by QDD, QET and classical ET models. The device was biased with $Vg = 0.8$ V and $Vd = 0.8$ V. The simulated density distributions are plotted at different positions of the channel. Fig. 4shows the electron density distributions perpendicular to the interface at the source end of the channel. The electron density distributions calculated from the QET and QDD models are almost identical in the inversion layers. Carrier heating due to the short channel effects results in the spread of electrons towards the bulk in simulations using the QET and ET models. As a result, the profiles between two models are almost identical at the bulk. The electron density distributions perpendicular to the interface at the drain end of the channel are shown in Fig. 5. The results clearly indicate that the quantum confinement effect is reduced by the enhanced diffusion towards the bulk due to the high electron temperature near the drain. The QET model allows simulations of quantum confinement transport with hot-carrier effects in MOSFETs.

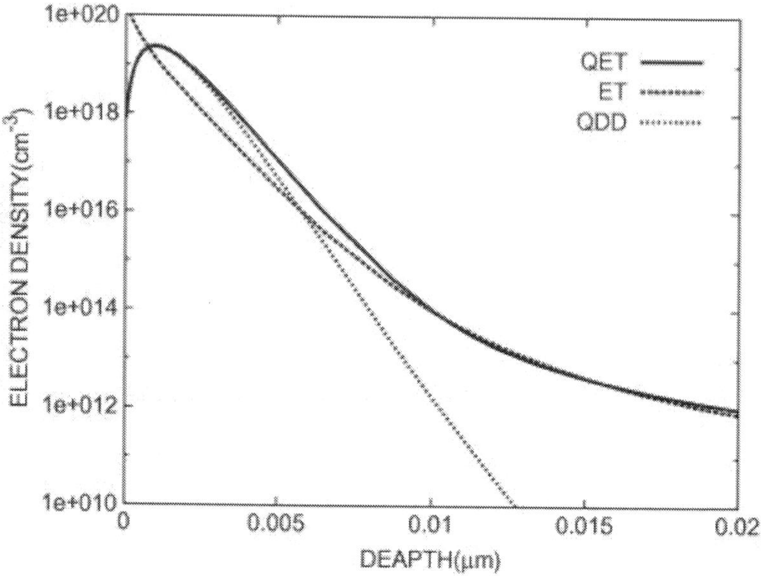

Figure 4. Electron density distributions perpendicular to the interface at the source end of the channel for a 35 nm MOSFET.

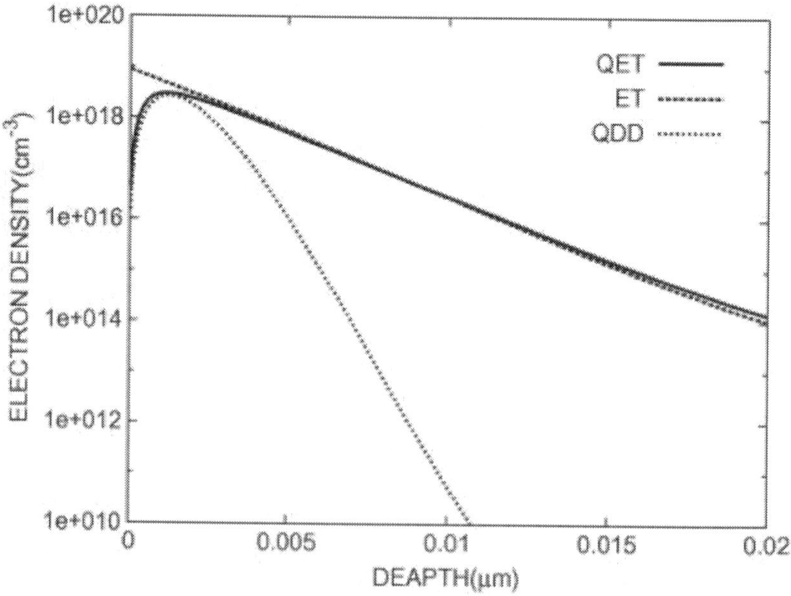

Figure 5. Electron density distributions perpendicular to the interface at the drain end of the channel for a 35 nm MOSFET.

Fig. 6 shows lateral profiles of electron temperature calculated by the QET, QCET, and ET models at the same gate voltage of 1.2 V. In Fig. 7, we compare the results calculated by the ET model at $Vg = 1.2$ V and the QET model at $Vg = 1.6$ V. The simulations are done at the same drain voltage of 0.8 V. The quantum corrected ET (QCET) model is a simplified QET model based on [4] with a temperature dependent mobility model (38). In the QCET model, the quantum correction to the energy density is neglected, and the carrier temperature in the current density is approximated by the lattice temperature [4]. As shown in Fig. 6, the QET model exhibits a sharper distribution of electron temperature at the lateral direction, when compared to that calculated by the classical ET model. The electron temperature calculated by the QCET model is further increased. This difference is caused by the threshold voltage shift due to the quantum confinement transport in the channel. Therefore, as shown in Fig. 7, the shape of electron temperature distributions calculated by the QET model at $Vg = 1.6$ V is close to that obtained by the ET model at $Vg = 1.2$ V. In Fig. 8, we present the x-component of the current densities calculated by the QET and ET models. The results verify that the magnitude of the current density calculated by the QET model at $Vg = 1.6$ V corresponds to that calculated by the ET model at $Vg = 1.2$ V.

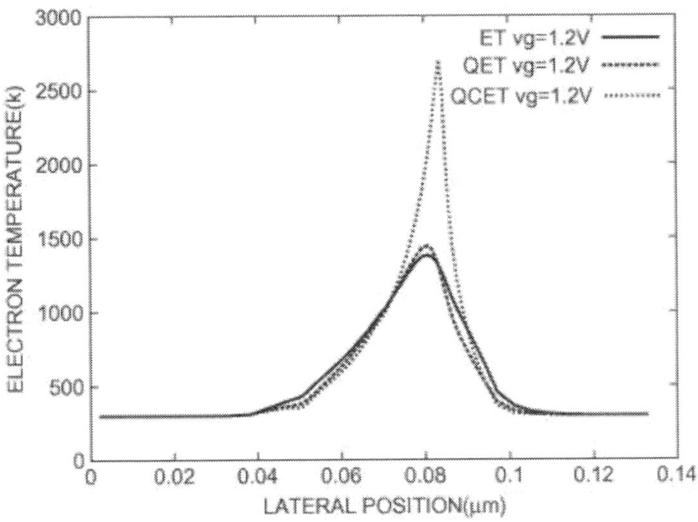

Figure 6. Lateral profiles of electron temperature distributions calculated by ET (solid line), QET, and QCET models at the same drain bias of $Vd = 0.8$ V and the same gate bias of $Vg = 1.2$ V.

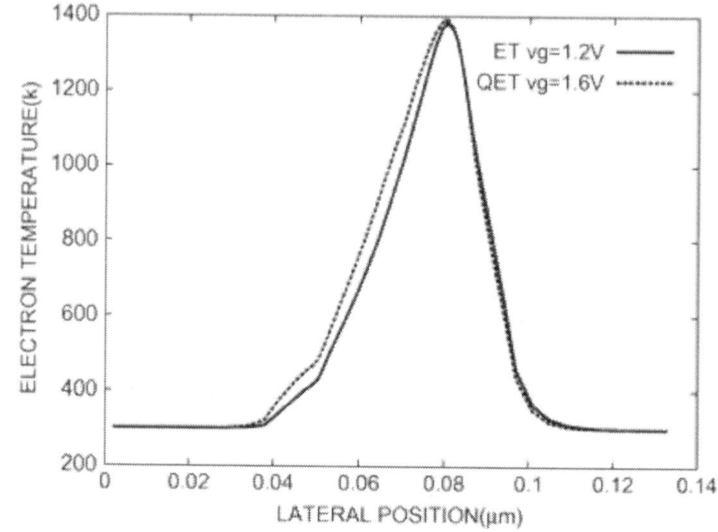

Figure 7. Lateral profiles of electron temperature distributions calculated by ET (solid line) and QET (dotted line) models at the same drain bias of $Vd = 0.8$ V. ET model at $Vg = 1.2$ V, QET model at $Vg = 1.6$ V.

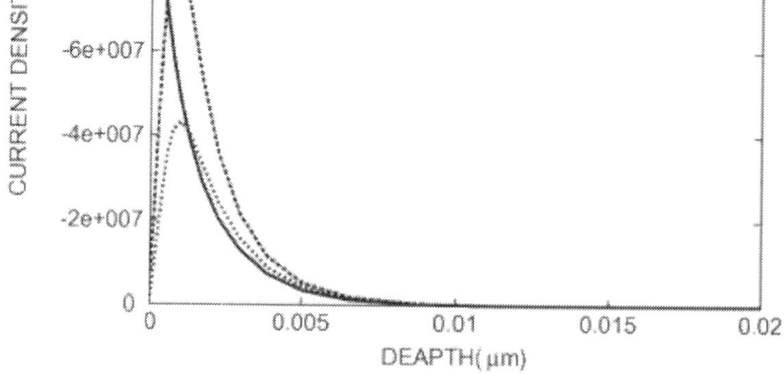

Figure 8. x-Component of current densities perpendicular to the interface for a 35 nm MOSFET. ET model at $Vg = 1.2$ V, QET model at $Vg = 1.2$ V and $Vg = 1.6$ V.

CONCLUSION

A four-moments QET model has been derived by using a diffusion scaling in the quantum hydrodynamic model. Space discretization of the four-moments QET model has been performed by a new set of unknown variables. Numerical schemes result in a consistent generalization of the Scharfetter-Gummel type scheme to the QET equations. We can enhance the robustness of the iterative solution method by introducing a relaxation method. The QET model allows simulations of quantum confinement transport with hot-carrier effects in scaled MOSFETs. The simulation results reveal the difference of electron temperature distributions between the QET and ET models due to the quantum confinement effects.

ACKNOWLEDGEMENT

The authors thank Dr. Shimada for numerical simulations.

REFERENCES

1. P. Degond, F. Méhats, C. Ringhofer, Quantum energy transport and drift diffusion models, J. Stat. Phys. 118 (2005) 625–667.
2. G. Allaire, A. Arnold, P. Degond, T.Y. Hou, Quantum Transport, Springer, 2008. pp. 144–152.
3. S. Jin, Y.-J. Park, H.-S. Min, Simulation of quantum effects in the nano-scale simiconductor device, J. Semi. Tech. Sci. 4 (2004) 32–38.
4. R.-C. Chen, J.-L. Liu, An accelerated monotone iterative method for the quantum-corrected energy transport model, J. Comp. Phys. 204 (2005) 131–156.
5. T. Grasser, T.-W. Tang, H. Kosina, S. Selberherr, A review of hydrodynamic and energy-transport models for semiconductor device simulation, IEEE Proc. 91 (2003) 251–274.
6. C.L. Gardner, The quantum hydrodynamic model for semiconductor devices, SIAM J. Appl. Math. 24 (1994) 409–427.
7. S.-C. Lee, T.-W. Tang, Transport coefficients for a sililcon hydrodynamic model extracted from inhomogeneous monte-carlo calculations, Solid-State Elec. 35 (1992) 561–569.

8. A. Bringer, G. Schön, Extended moment equations for electron transport in semiconducting submicron structures, J. Appl. Phys. 64 (1988) 2447–2455.
9. R. Thoma, A. Emunds, B. Meinerzhagen, H.J. Peifer, W.L. Engl, Hydrodynamic equations for semiconductors with nonparabolic band structure, IEEE Trans. Electron Devices 38 (1991) 1343–1353.
10. M. Gritsch, H. Kosina, T. Grasser, S. Selberherr, Revision of the standard hydrodynamic transport model for SOI simulation, IEEE Trans. Electron Devices 49 (2002) 1814–1820.
11. M.G. Ancona, G.J. Iafrate, Quantum correction to the equation of state of an electron gas in a semiconductor, Phys. Rev. B 39 (1989) 9536–9540.
12. E. Wigner, On the quantum correction for thermodynamic equailibrium, Phys. Rev. 40 (1932) 749–759.
13. S. Odanaka, Multidimensional discretization of the stationary quantum drift-diffusion model for ultrasmall MOSFET structures, IEEE Trans. CAD ICAS 23 (2004) 837–842.
14. B. Meinerzhangen, W.-L. Engl, The influence of the thermal equilibrium approximation on the accuracy of classical two-dimensional numerical modeling of silicon submicrometer MOS transistors, IEEE Trans. Electron Devices 35 (1988) 689–697.
15. M. Rudan, F. Odeh, Multi-dimensional discretization scheme for the hydrodynamic model of semiconductor devices, COMPEL 5 (3) (1986) 149–183.
16. T.-W. Tang, Extension of the Scharfetter–Gummel algorithm to the energy balance equation, IEEE Trans. Electron Devices ED-31 (1984) 1912–1914.
17. A. Forghieri, R. Guerrieri, P. Ciampolini, A. Gnudi, M. Rudan, G. Baccarani, A new discretization strategy of the semiconductor equations comprising momentum and energy balance, IEEE Trans. Comput. Aided Des. 7 (1988) 231–242.
18. D. Chen, E-C. Kan, U. Ravaioli, C-W. Shu, R-W. Dutton, An improved energy transport model including nonparabolicity and non-Maxwellian distribution effects, IEEE Electron Device Lett. 13 (1992) 26–28.
19. S. Odanaka, A high-resolution method for quantum confinement transport simulations in MOSFETs, IEEE Trans. CAD ICAS 26 (2007) 80–85.
20. H.K. Gummel, A self-consistent iterative scheme for one-dimensional steady state transistor calculations, IEEE Trans. Electron Devices 11 (1964) 455–465.
21. C. de Falco, E. Gatti, A.L. Lacaita, R. Sacco, Quantum-corrected drift-diffusion models for transport in semiconductor devices, J. Comput. Phys. 204 (2005) 533–561.

CITATION

Shohiro Sho, Shinji Odanaka, A quantum energy transport model for semiconductor device simulation, Journal of Computational Physics, Volume 235, 15 February 2013, Pages 486-496, ISSN 0021-9991, http://dx.doi.org/10.1016/j.jcp.2012.10.051.

CHAPTER 3

Modeling and Simulation of Dynamics and Noise of Semiconductor Lasers Under Ntsc Modulation for Use in the Catv Technology

Alaa Mahmoud[1], Safwat W.Z. Mahmoud[2], Kamal Abdelhady[2]

[1] High Institute for Engineering and Technology, El-Minia, Egypt

[2] Department of Physics, Faculty of Science, Minia University, 61519 El-Minia, Egypt

ABSTRACT

We investigate the dynamics and noise of semiconductor lasers (SLs) subject to analog modulation with the frequency plan of the National Television Standards Committee (NTSC) for use in the community access television (CATV) systems. The investigations are done in both the time and frequency domains as important methodologies to upgrade the optical CATV systems. The study is based on the rate equation model of the laser. The modulation dynamics is classified into four distinct types according to the waveform of the modulated signal, and the frequency spectra of the relative intensity noise (RIN) of these types are characterized. We show that the laser emits continuous and regular period signals under weak modulation. When the modulation index exceeds 43%, i.e. the modulation current exceeds the bias level above threshold, the laser emits clipped signals superposed by relaxation oscillation. The increase in the modulation index by about 50% in channels beyond channel #24 makes the laser emitting clipped pulsed signals superposed by relaxation oscillation. The laser attains higher noise levels (~−108 dB/Hz) when it emits pulses, whereas the noise is lowest (~−170 dB/Hz) when the signal is continuous.

INTRODUCTION

SLs are characterized by very attractive features that make them the most important light sources for many applications. One of the most important applications is the optical fiber communication network (Hashemi, 2012), which substituted the copper-based telecommunication networks to keep up with the increasing speed of information and the demand of subscribers on wideband CATV transport systems. Using optical fiber improves the picture quality and increases the CATV system reliability (Lu, 2010a). For the distribution of CATV channels over the frequency spectrum, as well as for the channel bandwidth, there are several standards, including the NTSC, Sequential Color with Memory (SECAM) and Phase Alternating Line (PAL) (Water, 2005).

The NTSC system was endorsed in 1941 for black and white broadcasting and was modified in 1953 to allow for color television broadcasting (Brillant, 2008). Fig. 1 depicts the NTSC frequency plan for the CATV signal transport incorporating a return path (up-stream), which includes sending signals back from the user to the head-end such as in the video on demand technology. The allocated band for the up-stream ranges between 5 and 42 MHz, and is divided into channels with bandwidth ranging between 1 and 3 MHz. There is a guard band between the up-stream and the CATV band (down-stream). The down-stream band spans 50–550 MHz for analog signals and spans 550–750 MHz for digital signals, which could be expanded even to 1 GHz (Mahmoud, 2011).Fig. 2 shows the frequency spectrum of one NTSC channel. The nominal channel spacing is 6 MHz, except for a 4 MHz frequency gap between channels # 4 and 5. The audio carrier is set to be 4.5 MHz higher than the video carrier, which is placed 1.25 MHz higher than the lower boundary of the channel. The video carrier for channels 4 and 5 are located 0.75 MHz below the 6 MHz multiples. The color subcarrier is placed approximately 3.58 MHz above the visual carrier. It contains the picture's color and is phase modulated (HP Company, 1994 and Mahmoud, 2011).

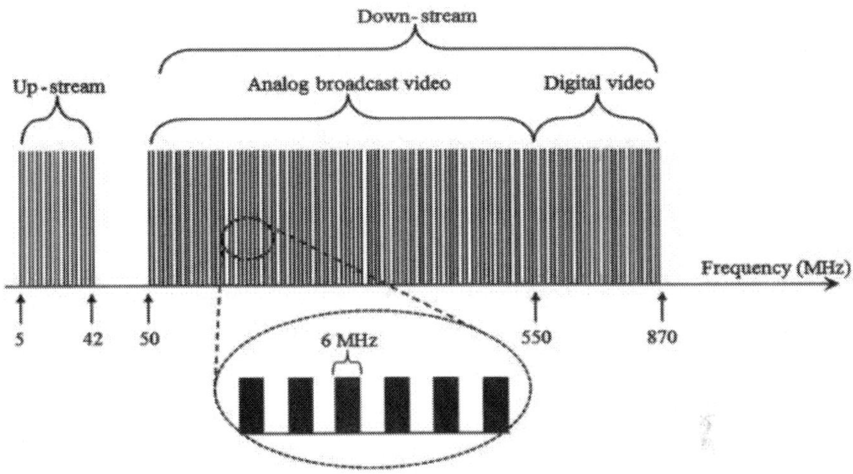

Figure 1. NTSC frequency plan for CATV signal transport, showing both up and down-stream with their frequency bands (Lu, 2010a).

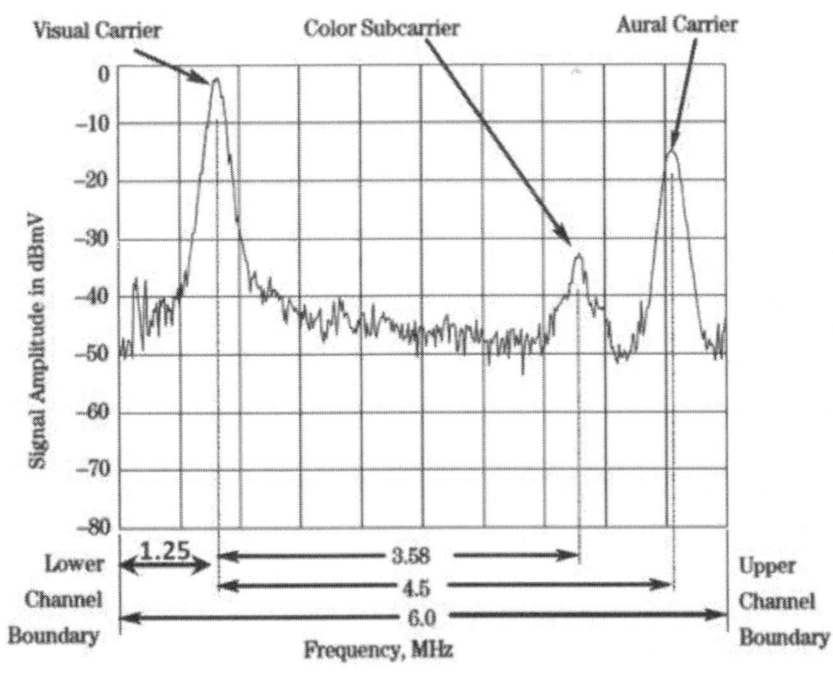

Figure 2. NTSC spectrum structure for one channel, showing the visual video, color, and audio carriers (HP Company, 1994).

In order to achieve high quality performance of fiber optical CATV systems using analog NTSC standards, a high carrier-to-noise ratio (CNR) is required, which is one of the most important radio frequency (RF) parameters (Chung and Jacobs, 1992, Rainal, 1995, Tzeng et al., 2005 and Lu et al., 2010b). Cost-3effective fiber communication systems adopt direct modulation of SL. However, the directly-modulated fiber systems are limited by the nonlinear phenomena of SL (Lu et al., 2010b). The SL dynamics are influenced by the coupling nature of the injected charge carriers with the emitted photons in the active region. During the laser transients, this coupling manifests as time delay of the photon emission and damped relaxation oscillations (Agrawal and Dutta, 1993). Under analog modulation, longer turn-on delay time and setting time of the relaxation oscillations (time instant at which the oscillations die) result in complicated nonlinear dynamics (Hori et al., 1988, Ahmed and El-Lafi, 2008a, Ahmed, 2008, Neo, 2001 and Ahmed et al., 2012a).

On the other hand, the power level of the SL has random fluctuations induced by spontaneous emission (Coldren and Corzine, 1995, Obarski and Hale, 1999, Abdulrhmann et al., 2003, Ahmed, 2004, Water, 2005, Hui and Sullivan, 2009, Ahmed et al., 2012b and Ahmed et al., 2013). Above the lasing threshold, SLs mostly emit stimulated emission and also a small amount of spontaneous emission. Since the spontaneous emission is a random process, a small part of the emitted photons may coincide with the wavelength and the direction of the stimulated emission photons and produce variations in the laser output (Agilent Technologies, 2008). The nearly absence of thermal noise in lasers makes the spontaneous emission noise the primary source of intensity noise (Lien, 2002). The SL noise is commonly used by RIN, which is defined in units of dB/Hz (Water, 2005) and strongly affects the CNR (Burden et al., 1981,Petermann, 1988, Saleh, 1989, Darcie and Bodeep, 1990, Chung and Jacobs, 1991,Phillips and Darcie, 1991, Frigo et al., 1993, Lai and Conradi, 1997, Movassaghi et al., 1998, Lu and Lee, 2000, Agilent Technologies, 2000, Agrawal, 2002, Water, 2005, Lu et al., 2010b and Mahmoud, 2011). Movassaghi et al. (1998) showed that selection of direct modulation may lead to a high intensity noise, and for optimal CNR, analog lasers should be chosen on the basis of RIN. The modulation index has to be as high as possible to reduce the negative impact of laser RIN (Darcie and Bodeep, 1990 and Water, 2005). However, if the modulation index exceeds the threshold current of the laser, occasionally the input current will drop below the laser threshold current, which results in nearly zero output power and shuts the laser off. This phenomenon is called clipping (Frigo et al., 1993 and Lai and Conradi, 1997), which is the fundamental limiting factor in analog communication (Water, 2005). Therefore, the modulation index should be limited to avoid this nonlinearity (Saleh, 1989, Chung and Jacobs, 1991, Phillips and Darcie, 1991 and Water, 2005). Therefore, it important to examine the influence of the modulation parameters; namely, the CATV modulation frequency (CATV channels) and the modulation index on the performance of SL in CATV systems and to characterize the associated noise. This would help system designers to choose the optimal operation conditions that yield regular dynamics or to avoid operation with signal clipping, distortion and noise enhancement.

In this paper, we introduce comprehensive simulation of SL dynamics and the associated intensity noise under direct modulation with the analog down-stream frequency band of the NTSC frequency plan. The studies were based on the rate equation model of SLs (Petermann, 1988). The rate equations are linearized by a small-signal approximation to determine the small-signal modulation response (Agrawal, 2002). A large-signal analysis of the NTSC modulation of SL is performed using OptiSystem software. Basing on the signal characteristics in the time and frequency domains, we classify the modulated laser signal into four types. These types are "sinusoidal signal (SS)", "periodic signal (PS)", "clipped signals with relaxation oscillations (CSRO)" and "clipped pulse with relaxation oscillation (CPRO)". The classification is done basing on the time characteristics of the modulated laser signal as well as its frequency characteristics. The noise properties of each type are determined by the frequency spectrum of RIN. The study is applied to distributed feedback (DFB) SLs due to their importance in the optical CATV systems (Movassaghi et al., 1998, Lu and Lee, 2000 and Water, 2005).

THEORETICAL AND CALCULATION MODEL

The dynamic behavior and modulation characteristics of SLs are modeled by the following pair of rate equations of the photon density *S* and injected carrier density *N* for a single-mode laser modulated by current *I(t)* (Corvini and Koch, 1987)

$$\frac{dS(t)}{dt}=\frac{\Gamma g_o(N(t)-N_o)}{1+\varepsilon S(t)}S(t)-\frac{S(t)}{\tau_p}+\frac{\Gamma\beta N(t)}{\tau_c}+F_s(t) \tag{1}$$

$$\frac{dN(t)}{dt}=\frac{I(t)}{eV}-\frac{N(t)}{\tau_c}-\frac{g_o(N(t)-N_o)}{1+\varepsilon S(t)}S(t)+F_N(t) \tag{2}$$

The output power *P(t)* is related to the emitted photon density *S(t)* through the relationship:

$$P(t)=\frac{V\eta h\nu}{2\Gamma\tau_p}S(t) \tag{3}$$

In the above equations, υ is the optical frequency, *h* is the Planck's constant, *η* is the deferential quantum efficiency, *Γ* is the confinement factor, *No* is carrier density at transparency, *β* is the fraction of spontaneous emission noise coupled into the lasing mode, *go* is the differential gain coefficient, ε is the nonlinear gain compression factor (gain saturation coefficient), *τp* is the photon lifetime, *τc* is the carrier lifetime, *V* is the active layer volume and *α* is the linewidth enhancement factor. The used values of these laser parameters are defined in Table 1. The injection current *I(t)* is given by:

$$I(t)=I_b+I_m\times\psi_m(t) \tag{4}$$

where *Ib* is the bias current, *Im* is the modulation current, and $\psi_m(t)$ represents the shape of the current signal. For single-tone modulation:

$$\psi_m(t) = A\sin(2\pi f_m t + \theta_j) \tag{5}$$

where *A* is the amplitude, *fm,* is the CATV modulation frequency (50–550 MHz) and θis the phase. The modulation index (*m*) is defined as (Brillant, 2008)

$$m\% = \frac{A \times I_m}{I_b} \times 100 \tag{6}$$

Table 1. Typical values of the parameters of a DFB laser employed in the calculations.

Symbol	Definition	Value	Unit
λ	Wavelength	1550	nm
V	Active layer volume	1.5×10^{-10}	cm^3
η	Quantum efficiency	0.4	
g_o	Differential gain coefficient	2.5×10^{-16}	cm^2
N_o	Carrier density at transparency	1×10^{18}	cm^{-3}
Γ	Mode confinement factor	0.4	
τ_c	Carrier lifetime	1×10^{-9}	S
τ_p	Photon lifetime	3×10^{-12}	S
β	Spontaneous emission factor	3×10^{-5}	
ε	Gain compression coefficient	1×10^{-17}	cm^3
α	Linewidth enhancement factor	5	
I_{th}	Threshold current	33.45	mA
I_b	Bias current	60	mA

The measurements is performed by setting *Ib* and *A* at a fixed values and varying *Im*

The last terms *FS(t)* and *FN(t)* in rate Eqs. (1) and (2) are Langevin noise sources with zero mean values, and are added to the equations to account for intrinsic fluctuations of the laser (Ahmed et al., 2001). The spectra of RIN are originally defined as the Fourier transform of the auto-correlation functions:

$$RIN = \frac{1}{\overline{P}^2} \int_0^{\infty} \delta P(t)\delta P(t+\tau)e^{j\omega\tau}d\tau \tag{7}$$

where ω is the Fourier angular frequency. Then, RIN calculated over a long time period *T* from the equation (Ahmed et al., 2001)

$$RIN = \frac{1}{\overline{P}^2}\left\{\frac{1}{T}\int_0^T\left[\int_0^{\infty}\delta P(t)\delta P(t+\tau)e^{j\omega\tau}d\tau\right]dt\right\} = \frac{1}{\overline{P}^2}\left\{\frac{1}{T}\left|\int_0^T \delta P(\tau)e^{-j\omega\tau}d\tau\right|^2\right\} \tag{8}$$

RESULT AND DISCUSSIONS

Light-current (L-I) characteristics

The most important characteristic of any laser diode to be measured is the amount of light it emits as current is injected into the device. This generates the output light versus input current curve, more commonly referred to as the (L-I) curve (Mobarhan, 1995). The light power *P* emitted from the laser is calculated from the emitted photon density *S(t)* via relationship (3). Fig. 3 plots the calculated L-I curve, the threshold current *Ith*, which is a very important parameter since it is strictly related to the power consumption of the laser, is gotten from the L-I curve by checking the critical current at which the output optical power starts an increasing linear behavior. For the used laser model, *Ith* = 33.45 mA and at *Ib* = 60 mA, the light output power is 4.24 mW (6.2 dBm). The regime below *Ith*corresponds to the spontaneous emission and is characterized by very low power. When*I* > *Ith*, the laser power abruptly increases with the little increase in the current *I* due to the stimulated emission.

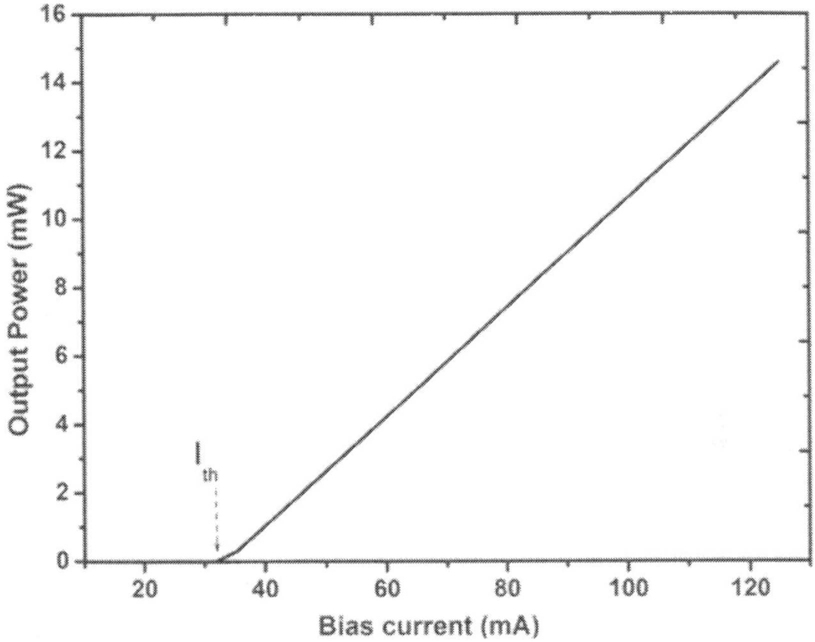

Figure 3. Output light vs. current injection for laser model.

Small-signal modulation transfer function

The small-signal modulation response is one of the most important useful parameters to evaluate the direct analog modulation performance of SLs. It defines the transfer function from current modulation to optical output power (Ahmed and El-Lafi, 2008b), and acts as an indicator for the possible modulation frequencies. The small-signal modulation response for various bias currents could provide a significant insight to the laser dynamics (Ahmed and El-Lafi, 2008b). This response is simulated by calculating the fast Fourier transformation (FFT) of the modulated signal at small values of *m*, and picking up the amplitude of the peak at the fundamental frequency of each channel. Fig. 4 plots the frequency spectra of the simulated modulation response *H (fm)* at very weak modulation corresponding to *m* = 3.5% when *Ib* = 40, 60, 90 and 150 mA. The figure shows that *H(fm)*exhibits a pronounced peak at a frequency *fm(peak)* ≈ *fr*. As numeric examples, *fr* = 2.4, 4.7, 6.8, 9.4 GHz when *Ib* = 40, 60, 90 and 150 mA. These spectra can be well described by the following normalized small-signal modulation response (Bowers, 1987)

$$H(f_m; Y, Z) = \frac{Z}{(j2\pi f_m)^2 + j2\pi f_m Y + Z} \quad (9)$$

where Y and Z are functions of the laser parameters and bias current

$$Y = \frac{g_o\overline{S}}{1+\varepsilon\overline{S}} + \frac{1}{\tau_c} - \frac{\Gamma g_0\left(\overline{N}-N_0\right)}{\left(1+\varepsilon\overline{S}\right)^2} + \frac{1}{\tau_p} \quad (10)$$

$$Z = \frac{g_o\overline{S}}{\left(1+\varepsilon\overline{S}\right)\tau_p} + \frac{(\beta-1)\Gamma g_0\left(\overline{N}-N_0\right)}{\tau_c\left(1+\varepsilon\overline{S}\right)^2} + \frac{1}{\tau_c\tau_p} \quad (11)$$

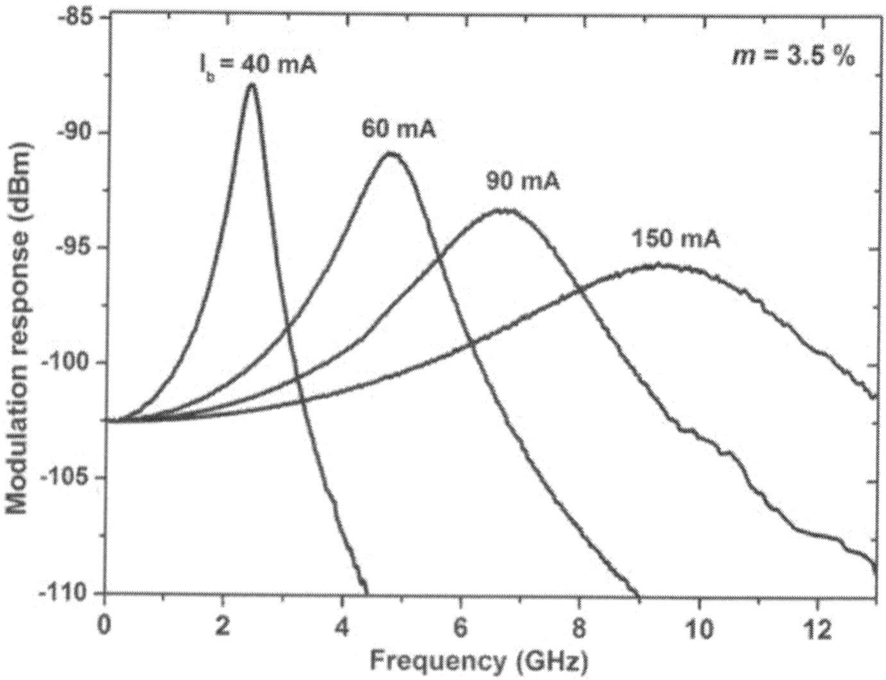

Figure 4. The spectrum of *H (fm)* when *Ib* = 40, 60, 90 and 150 mA. The response peak decreases and *H (fm)* broadens with the increase of *Ib*.

$\overline{N}$ and $\overline{S}$ denote the steady-state values of the carrier and photon densities corresponding to the bias current of the laser. The above analysis is obtained by linearizing the rate equations (1) and (2) under the common approximation of $I_m << I_b$ of the small-signal analysis (Bowers, 1987).

Fig. 4 shows also that the increase in *Ib* is associated with shift of the response peak to higher modulation frequencies, decrease of the response peak, and broadening of the spectrum around the peak frequency. This spectrum can be understood as follows (Wu, 1995). When *fm* is much lower than *fr*, the flat response is because the injected carriers follow the change in the injection current. The response peak is because the charge carriers interact with the photons with phase synchronization, which results in the laser resonance. The declining part of *H (fm)* is because the phase of the photon field lags behind that of the injection current. As *fm* is increased beyond *fr*, the electron and photon fields tend to become more and more out of phase, resulting in damping of the relaxation oscillations and the shown monotonic decrease of *H (fm)*.

Characterization of the modulated laser signal

The time and frequency domain investigations of the deterministic laser dynamics under the NTSC modulations indicate 4 types of the modulated laser signal. The Langevin noise sources are dropped from rate equations (1) and (2) for such investigation. The characteristics of these signal types are illustrated in Fig. 5, Fig. 6, Fig. 7 and Fig. 8. The figures illustrate the time domain presentation of the signal, its (power *P* versus population inversion *N–Nth*) phase portrait, and the corresponding power spectrum.Fig. 5(a) and (b) correspond to the "SS" type which occupies the lower range of $m = 15\%$ with the modulation frequency $fm = 205.25$ MHz (Channel #12). Fig. 5(a) shows that the signal varies sinusoidally similar to the current *I* due to the weak modulation. The laser output varies regularly and symmetrically with the time variation. This type corresponds to a limit cycle in the phase portrait of Fig. 5(b) with positive correlation between *P* and*N–Nth*. The SS type, therefore, would yield lowest noise levels. The corresponding power spectrum is plotted in Fig. 5(c); it has a pronounced peak at $f = fm$ and other lower peaks at the higher harmonics.

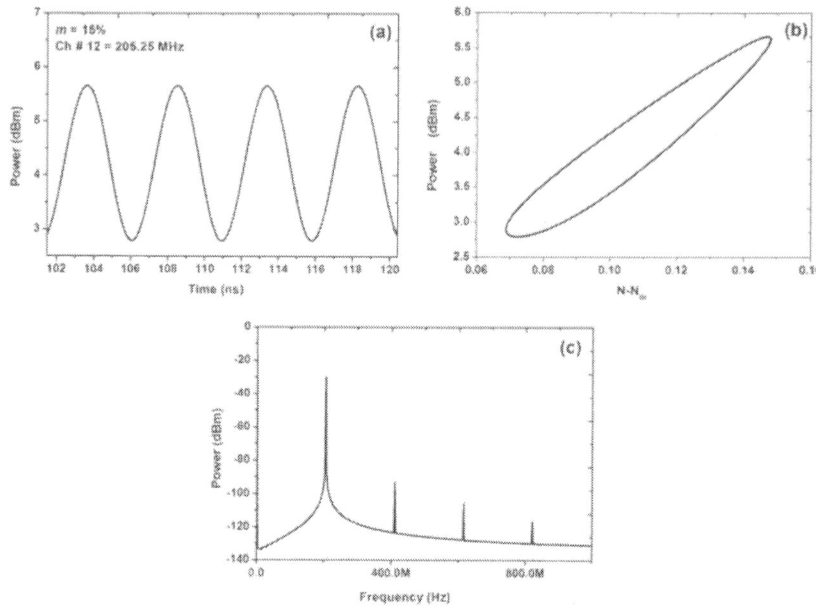

Figure 5. Typical characteristics of SS: (a) signal shape, (b) phase portrait, and (c) power spectrum, when m = 15% with Ch#12 (f_m = 205.25 MHz).

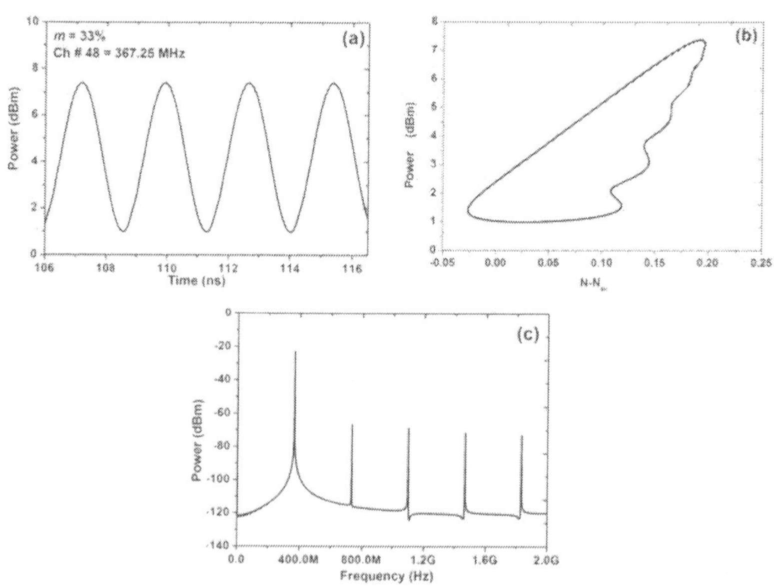

Figure 6. Typical characteristics of PS: (a) signal shape, (b) phase portrait, and (c) power spectrum, when m = 33% with Ch#48 (f_m = 367.25 MHz).

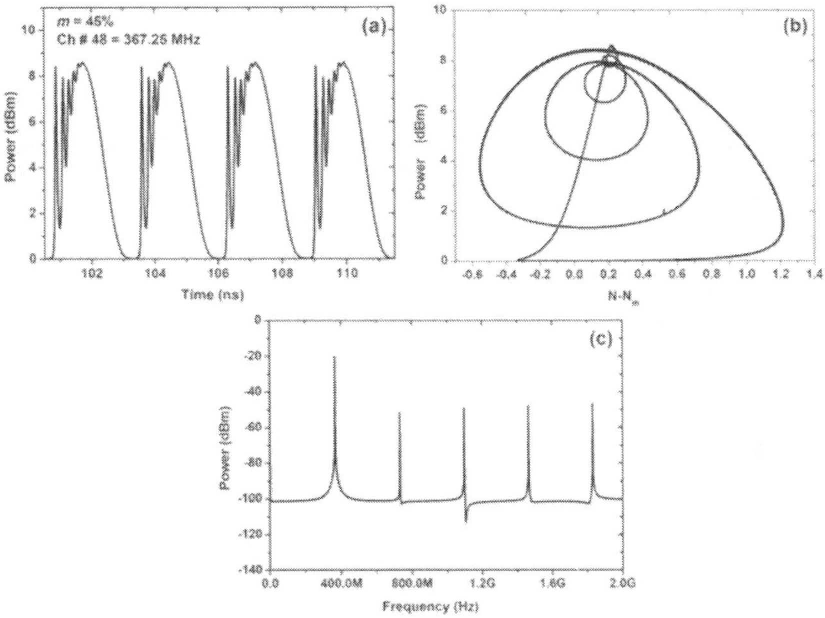

Figure 7. Typical characteristics of CSRO: (a) signal shape (b) phase portrait, and (c) power spectrum, when$m$ = 45% with Ch#48 (f_m = 367.25 MHz).

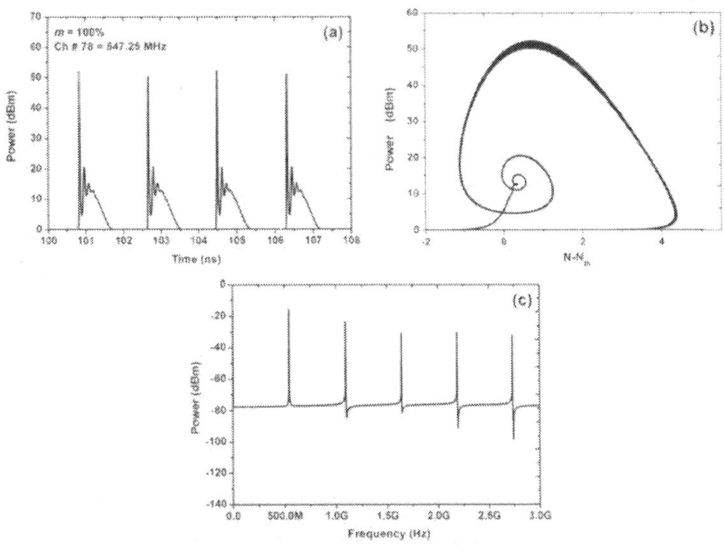

Figure 8. Typical characteristics of PPRO: (a) signal shape, (b) phase portrait, and (c) power spectrum, when$m$ = 100% with Ch#78 (f_m = 547.25 MHz).

Fig. 6(a) – (c) correspond to the “PS” type for which the signal is continuous but is not sinusoidal. This type corresponds to $m = 33\%$ with $fm = 367.25$ MHz (channel #48).Fig. 6(a) shows that the signal varies continuously and regularly but not sinusoidally because of the increase of m from 15% of the SS type to 33%. The asymmetry of the signal is reflected in the phase portrait of Fig. 6(b) which indicates a single loop attractor with its bottom being wider than its top. The corresponding power spectrum is plotted inFig. 6(c), and has a pronounced peak at $f = fm$ and other lower peaks at the higher harmonic frequencies.

Fig. 7(a) – (c) characterize another type of the periodic signals but with clipping superposition by relaxation oscillation “CSRO”. This type is the development of the PS type when m increases. The figures correspond to $m = 45\%$ with the same channel frequency used in Fig. 6. Fig. 7(a) shows that the signal varies continuously, but clipped and superposed by sub-peaks from relaxation oscillations in each period. We can interpret these characteristics as follows. The relatively large value of m exceeds the threshold current of the laser, consequently the input current drops below the laser threshold current, which then results in nearly zero output optical power and therefore the signal is clipped. Also, when $fm \ll fr$, the period Tm is much longer than the setting time of the relaxation oscillations, therefore, with the instantaneous rise up of $I(t)$ the cycle duration is long enough to build up the relaxation oscillations in the signal. The multiple relaxation oscillation peaks are seen as multiple loops with different sizes in the phase portrait of Fig. 7(b). The corresponding power spectrum is plotted in Fig. 7 (c), which has a pronounced peak at $= fm$ and other lower peaks at the higher harmonics.

The last and noisiest investigated type for modulation dynamics of the analog NTSC system is “CPRO” which follows the CSRO type with the increase of m. This type is shown in Fig. 8(a) and (b) in which the laser emits clipped periodic pulses superposed with a sub-peak stemming from the relaxation oscillations. This figure corresponds to$m = 100\%$ with $fm = 547.25$ MHz (channel #78). Fig. 8(a) displays a clipped signal in the form of short and strong pulse followed by a relatively weak sub-peak during each period. In the phase portrait of Fig. 8(b), the overshoot peak in Fig. 8(a) is seen as a large loop attractor while the lower two peaks are seen as smaller loops. The power spectrum is plotted in Fig. 8(c) shows strongest peak at *fm* and weaker peaks at the higher harmonics.

(Modulation index versus channel frequency) diagram of modulation dynamics

In this section, the operating regions of the types of the SL signal under NTSC modulation are explored. The results help to determine the modulation parameters of each type at a given bias current. For such a purpose, *m* and the modulation frequency *fm*corresponding to each of the investigated four types are determined. The relevant ranges are *m* = 5% – 120%and *fm* = 55.25 MHz (channel #2) – 547.25 MHz (channel #78) in order to cover the analog down-stream frequency band. The obtained results are plotted in the diagram of Fig. 9.

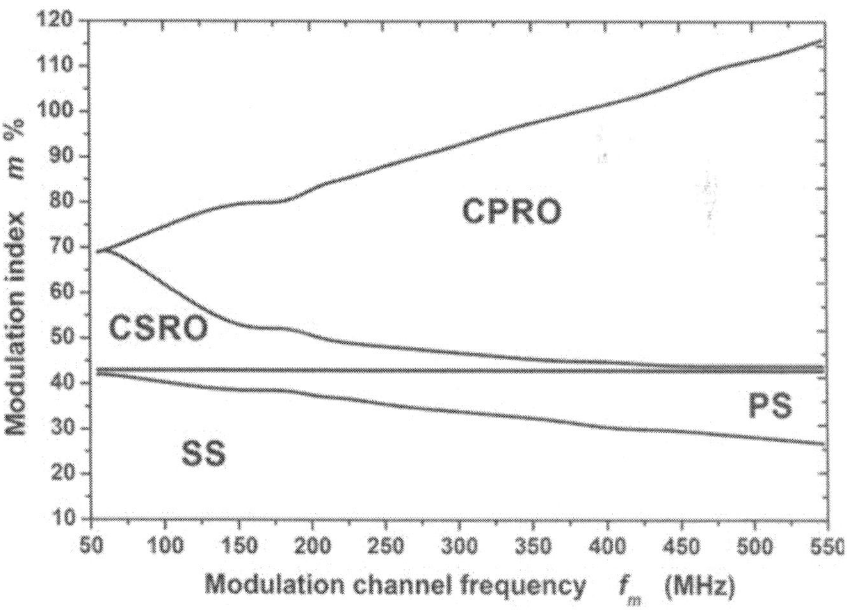

Figure 9. Two-dimensional diagram of *m* and channel frequency. The four dynamic types are.allocated as function of *m* and *fm*.

The modulation parameters (*m*, *fm*) that correspond to each type are decided by examining the time domain presentation of the signal and the corresponding power spectrum, as discussed above. Fig. 9 shows that the region of the SS type corresponds to $m \leq 42\%$ when $fm \leq 67.25$ MHz (channel #4), $m \leq 37\%$ when $fm \leq 223.25$ MHz (channel #24), $m \leq 35\%$ when 259.25NHz (channel #30) $\leq fm \leq 475.25$ MHz (channel #66), and $m \leq 27\%$ when $fm \leq 547.25$ MHz (channel #78). The PS type appears when $m \leq 43\%$ regardless the value of fm. The region of the CSRO type appears when the m slightly exceeds 43%, i.e. $Im > Ib–Ith$, as a threshold clipping condition, and its upper m border decreases with the increase in fm when $fm \leq 439.25$ MHz (channel #60) and is kept constant when $fm \leq 547.25$ MHz (channel #78). The CPRO appears when the fm reaches 55.25 MHz (channel #2) and when $m = 69\%$. The region of the CPRO then broadens with the increase in fm; it extends between $m = 50\%$ and 85% when $fm = 223.25$ MHz (channel #24), and between $m = 45\%$ and 116% when $fm = 547.25$ MHz (channel #78).

Noise characteristics of the modulated types of the SL signal

As shown above, the nonlinear effects in SLs induce inconsistency between the modulated laser signal and the modulating current signal. These distortions in the modulated signal may enhance the intrinsic noise level of the SL and deteriorate the coherency of the laser. In order to investigate the noise properties of the modulated laser, the Langevin noise sources in Eqs. (1) and (2) are taken into account. The noise is characterized in terms of the spectral characteristic of RIN and its low frequency level, LF-RIN.

Fig. 10 displays the frequency spectrum of RIN for the four types of the modulated signal.Fig. 10 (a) characterizes RIN of the SS type when $m = 15\%$ with $fm = 205.25$ MHz (channel #12). In addition to the characteristic peak at the relaxation frequency fr, the RIN spectrum has sharp peaks at fm and its higher harmonics. The low-frequency noise is as low as LF-RIN ~ - 170 dB/Hz, which is manifestation of the high degree of uniformity of the signal. The RIN spectrum characterizing the PS type is plotted in Fig. 10(b), it corresponds to $m = 33\%$ and $fm = 367.25$ MHz (channel #48). The LF-RIN level is comparable to that of the SS type. The RIN spectrum exhibits also peaks at fm and its higher harmonics. The RIN spectrum of the CSRO type is illustrated in Fig. 10(c) when the $m = 45\%$

with the same channel number (*fm*). The spectrum is higher than those of the SS and PS types; LF-RIN increases to ~ −160 dB/Hz due to the increase in the irregularity of the signal. The high-frequency part is characterized by higher peaks at *fr* as well as at *fm* and its multiples. Fig. 10(d) plots a typical RIN spectrum of the CPRO type, it is simulated when $m = 100\%$ and $fm = 547.25$ MHz (channel #78). The spectrum is characterized by *fm* and its higher harmonics. The LF-FN is flat with a level as high as LF- RIN ~ −108 dB/Hz, which indicates deterioration of the signal-to-noise ratio. These noise characteristics are in fit with the results predicted by Ahmed et al. (2012b) for analog modulation of AlGaAs laser diodes.

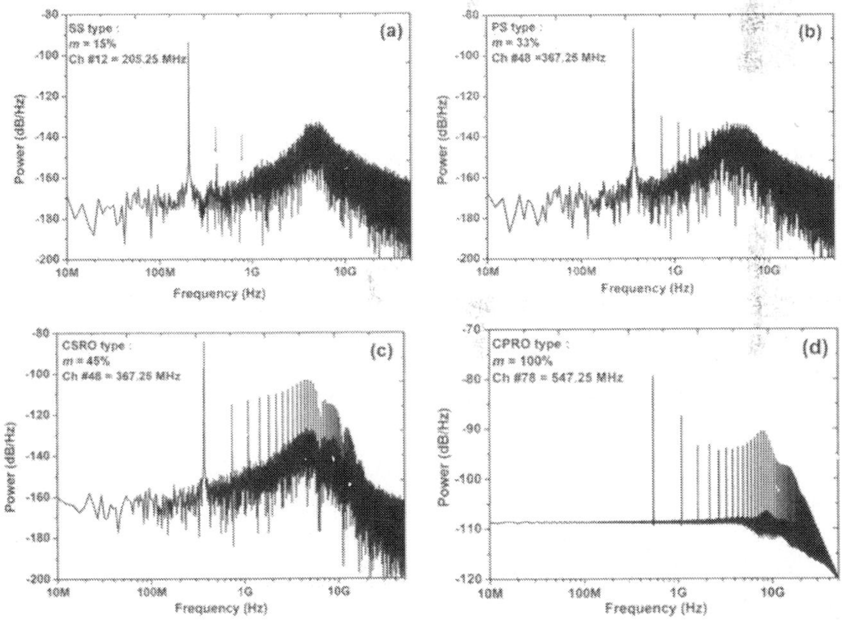

Figure 10. Simulated spectra of RIN at different modulation types and its modulation conditions.

CONCLUSIONS

We modeled and simulated the dynamics and noise of SL under direct modulation with the analog down-stream frequency band of the NTSC frequency plan. The modulation performance of the laser is investigated for the use in the CATV technology. The modulated signal of the laser is classified into four distinct types; namely, SS, PS, CSRO and CPRO. We can trace the following conclusions from the obtained results. The SS type corresponds to low values of *m*. The PS type follows the SS type when *m* increases, and its upper border of *m* is 43% regardless the value of *fm*. The further increase in *m*, i.e.*Im* > *Ib–Ith*, develops the PS type to the CSRO type. The CPRO type is a natural extend of CSRO with the increase in *m*. The region of the CPRO broadens with the increase in *fm*. The transitions between these types with the increase in *m* are almost associated with an increase in LF-RIN. The SS and PS types are characterized by minimum LF-RIN. The highest LF-RIN level, (−108 dB/Hz) is obtained under the modulation of CPRO. In general, when the laser emits pulses the noise spectra become highest, whereas these noise spectra become lowest when the laser signal is continuous.

REFERENCES

1. Abdulrhmann S, Ahmed M, Yamada M. New model of analysis of semiconductor laser dynamics under strong optical feedback in fiber communication systems. J SPIE 2003;4986:490e501.
2. Agrawal GP. Fiber-optic communication systems. New York. 2002. Agrawal GP, Dutta NK. Semiconductor lasers. New York. 2nd ed. 1993.
3. Ahmed M. Numerical approach to field fluctuations and spectral lineshape in InGaAsP laser diodes. Int J Numer Model 2004;17:147e63.
4. Ahmed M. Spectral lineshape and noise of semiconductor lasers under analog intensity modulation. J Phys D 2008;41(17):175104e13. Ahmed M, El-Lafi A. Analysis of small-signal intensity modulation of semiconductor lasers taking account of gain suppression. Pram J Phys 2008a;71:115e90.
5. Ahmed M, Ellafi A. Large-signal analysis of analog intensity modulation semiconductor lasers. J Opt Laser Tech 2008b;40:809e19.
6. Ahmed M, Yamada M, Saito M. Numerical modeling of intensity and phase noise in semiconductor lasers. J Quan Elect. IEEE 2001;37(12):1600e10.
7. Ahmed M, Mahmoud SWZ, Mahmoud A. Influence of pseudorandom bit format on the direct modulation performance of semiconductor lasers. Pram J Phys 2012a;79:1443e56.

8. Ahmed M, El-Sayed NZ, Ibrahim H. Chaos and noise control by current modulation in semiconductor lasers subject to optical feedback. J Eur Phys J D 2012b;66:141.
9. Ahmed M, Mahmoud SWZ, Mahmoud A. Comparative study on modulation dynamic characteristics of laser diodes using RZ and NRZ bit formats. Int. J Numer. Model 2013;27:138e52.
10. Bowers JE. High speed semiconductor laser design and performance. J Solid State Electron 1987;30:1e11.
11. Brillant A. Digital and analog fiber optic communications for CATV and FTTx applications. Bellingham: USA; 2008.
12. Burden RL, Faires JD, Reynolds AC. Numerical analysis. 2nd ed. Weber and Schmidt; 1981.
13. Chung CJ, Jacobs I. Simulation of the effects of laser clipping on the performance of AM SCM lightwave systems. J Int. J Numer. Model Phot Technol Lett 1991;3:1034e6.
14. Chung CJ, Jacobs I. Practical TV channel capacity of lightwave multichannel AM SCM systems limited by the threshold nonlinearity of laser diodes. J Phot Tech Lett IEEE 1992;4(3):289e92.
15. Coldren LA, Corzine SW. Diode lasers and photonics integrated circuits. New York: USA; 1995. HP Company. Cable television system measurement handbook (NTSC systems). USA: California; 1994.
16. Corvini PJ, Koch TL. Computer simulation of high bit rate optical fiber transmission using single frequency lasers. J Light Technol 1987;5:1591e5.
17. Darcie TE, Bodeep GE. Lightwave subcarrier CATV transmission systems. J IEEE Trans 1990;38:524e33.
18. Frigo NJ, Phillips MR, Bodeep GE. Clipping distortion in lightwave CATV systems: models, simulations, and measurements. J Light Tech 1993;11(1):138e46.
19. Hashemi SE. Relative intensity noise (RIN) in high-speed VCSELs for short reach communication. Master thesis. Swed Goteborg € 2012;1.
20. Hori Y, Serizawa H, Sato H. Chaos in a directly modulated semiconductor laser. J Opt Soc 1988;5:1128e113328.
21. Hui R, Sullivan MO. Fiber optic measurement techniques. USA. 2009.
22. Lai S, Conradi J. Theoretical and experimental analysis of clipping-induced Impulsive noise in AMeVSB subcarrier multiplexed lightwave systems. J Light Tech 1997;15(1):20e30.
23. Lien Y. Intensity dynamics of a slow-inversion laser. Netherlands: Doctor Thesis; 2002.
24. Lu HH. Fiber broadband network systems. Product note. Taiwan. 2010.
25. Lu HH, Lee CT. Long-distance transmission of directly modulated 1550 nm Fiber optical CATV transmission systems. Master thesis. Taiwan 2000:9.

26. Lu HH, Chang CH, Peng PC. In: Pal Bishnu, editor. Improvement scheme for directly modulated fiber optical CATV system performances7619; 2010. p. 82e4.
27. Mahmoud RM. Color spaces analysis for luminance & chrominance signals AS NTSC-TV system. J Eng. Dev 2011;1813:123e34.
28. Mobarhan KS. Test and characterization of laser diodes: determination of principal parameters. Doctoral thesis. USA: Newport Corporation. 1995. Movassaghi M, Jackson MK, Smith VM. DFB laser RIN degradation in CATV lightwave transmission. J Lasers Electro-Optics IEEE 1998;2:295e6.
29. Neo HB. Analysis of relative intensity noise and simulation of vertical-cavity surface-emitting lasers (Master Thesis). Australia: University of Queensland; 2001.
30. Obarski GE, Hale PD. How to measure relative intensity noise in lasers. J Las Foc. Wor 1999;35(5):273e7.
31. Petermann K. Laser diode modulation and noise. Dordrecht: Kluwer Academic; 1988.
32. Phillips MR, Darcie TE. Numerical simulation of clipping induced distortion in analog lightwave systems. J IEEE Phot Tech Lett 1991;3(12):1153e5.
33. Rainal AJ. Distortion spectrum of laser intensity modulation. J IEEE Trans 1995;43(11):1644e52.
34. Saleh AA. Fundamental limit on number of channels in subcarrier multiplexed lightwave CATV system. J Electron. Lett 1989;25(12):776e7.
35. Agilent Technologies. Lightwave signal analyzers measure relative intensity noise. Product Note 2000:71400e1. U.S.A.
36. Agilent Technologies. Digital communication analyzer (DCA), measure relative intensity noise (RIN). Product Note 2008:86100e7. U.S.A.
37. Tzeng SJ, Lu HHL, Chang CH, Peng PC. Employing split-band technique and optical SSB fiber to improve directly modulated fiber optical CATV system performances. J ICIE Elect Expr 2005;2(11):344e84.
38. Water MV. Low-cost CATV transmission in fiber-to-the-Home networks. Netherlands: Master thesis; 2005.
39. Wu CY. Analysis of high-speed modulation of semiconductor lasers by electron heating. Master thesis. Canada. 1995.

CITATION

Alaa Mahmoud, Safwat W.Z. Mahmoud, Kamal Abdelhady, Modeling and simulation of dynamics and noise of semiconductor lasers under NTSC modulation for use in the CATV technology, Beni-Suef University Journal of Basic and Applied Sciences, Volume 4, Issue 2, June 2015, Pages 99-108, ISSN 2314-8535, http://dx.doi.org/10.1016/j.bjbas.2015.05.002.

CHAPTER 4

Advances in Atomic Physics: Four Decades of Contribution of the Cairo University – Atomic Physics Group

Tharwat M. El-Sherbini,

Physics Department, Faculty of Science, Cairo University, Giza, Egypt

ABSTRACT

In this review article, important developments in the field of atomic physics are highlighted and linked to research works the author was involved in himself as a leader of the Cairo University – Atomic Physics Group. Starting from the late 1960s – when the author first engaged in research – an overview is provided of the milestones in the fascinating landscape of atomic physics.

INTRODUCTION

During the last decades, we witnessed a continuous development in the field of atomic physics that had direct impact on other fields of research such as astrophysics, plasma physics, controlled thermonuclear fusion, laser physics, and condensed matter physics.

The landscape is vast and cannot possibly be covered in one review article, but it would require a complete book. Therefore, I will confine myself to the research works I was involved in and those that have direct connections with the work I have done.

The review is structured around five main topics:

–Electron–atom collisions.
–Ion–atom collisions.
–Atomic structure calculations and X-ray lasers.
–Laser-induced breakdown spectroscopy (LIBS).
–Laser cooling and Bose–Einstein condensation.

ELECTRON–ATOM COLLISIONS

The physics of electron–atom collisions originated in 1930 by the work of Ramsauer and Kollath [1] and [2] on the total scattering cross-section of low energy electrons against noble gases, which contributed so much to the development of quantum theory. This work was followed by Tate and Smith [3] on inelastic total cross-sections for excitation of noble gases. Several well known physicists, e.g., Bleakney and Smith [4], Hughes and Rojansky [5], and Massey and Smith [6], at this period gave important contributions in the field of electron collision physics. The theory was developed by Stueckelberg [7], Landau [8], and Zener [9]. In 1952, Massey and Burhop's book [10] appeared on "Electronic and Ionic Impact phenomena," which provided the basis for any scientist who wants to start the work on the subject.

Multiple ionization of noble gases by low energy electrons (below 600 eV) has been studied extensively in mass spectrometers [3], [11] and [12]. However, total electron impact cross-sections were determined by Van der Wiel et al. [13] and El-Sherbini et al.[14] for the formation of singly and multiply charged ions of He, Ne, Ar, Kr, and Xe by fast electrons (2–16 keV). The ion selection was performed in a charge analyzer with 100% transmission, and consequently, it was possible to avoid the discrimination effects in the measurement of the relative abundances of the multiply charged ions. Therefore, the data were more reliable than those obtain in low transmission mass spectrometers. The ionization cross-section of large electron impact energies is given by

$$\frac{\sigma_{ni}}{4\pi a_0^2}\frac{E_{el}}{R} = M_{ni}^2 \ln E_{el} + C_{ni} \tag{1}$$

where σ_{ni} is the cross-section for formation of $n+$ ions, E_{el} is the electron energy corrected for relativistic effects, a_0 is the first Bohr radius, R is the Rydberg energy, M_{ni}^2, and C_{ni} are constants.

The constant M_{ni}^2 is given by

$$M_{ni}^2 = \int_{n+} \frac{df^{n+}}{dE}\frac{R}{E}dE \tag{2}$$

where df^{n+}/dE is the differential dipole oscillator strength for an ionization to n^+continuum at excitation energy E.

In 1970, an experiment was developed by van der Wiel [15], in which fast electrons (10 keV), scattered by He, Ne, and Ar are detected in coincidence with the ions formed (Fig. 1). It was possible from the measurements of the scattering intensity at small angles to calculate optical oscillator strengths. The differential scattering of fast electrons is given by Bethe et al. [16] (in au):

$$\sigma(\vartheta, E) = \frac{2}{E}\frac{k_n}{k_0}\frac{1}{K^2}\frac{df(K)}{dE} \tag{3}$$

where ϑ is the scattering angle, E the energy loss, k_0 and k_n arethe magnitudes of the momenta of the primary electron before and after collision, K is the magnitude of the momentum transfer ($K = k_0 - k_n$), and (df(K)/dE) is the generalized oscillator strength. This last quantity may be expanded in terms of K^2:

$$\frac{df(K)}{dE} = \frac{df}{dE} + aK^2 + bK^4 + \tag{4}$$

where $\frac{df}{dE}$ is the optical oscillator strength, as defined in the dipole approximation.

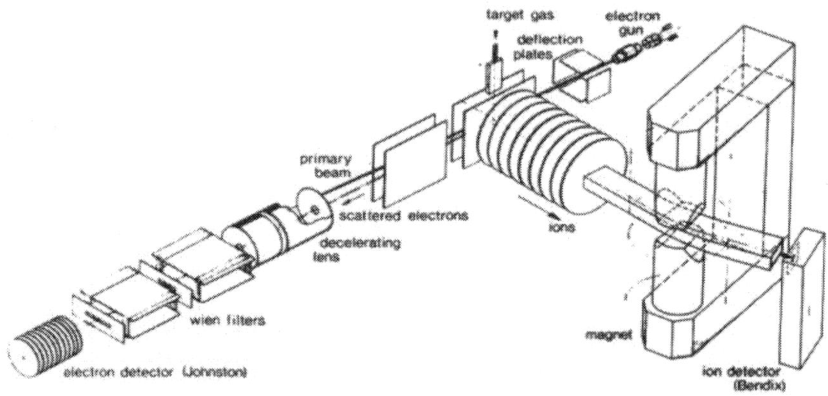

Figure 1. Schematic view of the scattered electron–ion coincidence apparatus.

The first table-top synchrotron

he work was closely connected to that where ion charge distribution is measured after irradiation of atoms with photons at a number of selected wavelengths [17] and [18]. However, the use of photon source is simulated by measuring the small-angle, inelastic scatting of 10 keV electrons in coincidence with the ions formed. The simulation is based on the fact that measured energy lost by the scattered electron in the coincident experiment corresponds to the photon energy absorbed in the photon experiments for the same process. Moreover, the incident electron energy of 10 keV is large compared to the energy losses studied ⩽400 eV, and also, the incident momentum (370 au) is much larger than the momentum transfer (⩽0.5 au). Under these conditions, the first Born approximation holds. By making use of the first Born approximation for inelastic electron scattering at small momentum transfer, the measured intensities of scattering were converted into optical oscillator strengths. Fig. 2 shows the block diagram of the electronic circuit, where signals from the ion and the electron detectors are measured in delayed coincidence. The true coincidences after being separated from the simultaneously registered accidental ones are stored in a data collector that drives the energy loss scanning. The number of true coincidences is recorded per number of ions of the charge state under consideration. This enables us to put spectra for different charge states on the same relative scale when knowing the relative abundances of the charge states at 10 keV electron impact energy. This technique combines the advantage of continuous variability of the energy transfer over a few hundred eV with that of a constant detection efficiency. As a result, oscillator-strength spectra over a wide energy range were obtained, which could be put on an absolute scale by normalization on an absolute photo-absorption value at only one energy. As far as the intensity is concerned, this method compares favorably with a possible alternative of charge analysis of ions formed by dispersed electron synchrotron radiation in a low density target (10^{-5} torr). This work was extended by El-Sherbini and van der Wiel [19] to measure oscillator strengths for multiple ionization in the outer and first inner shells of Kr and Xe (Fig. 3 and Fig. 4). Direct ejection of two N electrons below the $3d^9$ threshold is observed in the Kr^{2+} spectrum, which was found to be a characteristic of such transitions. The threshold for discrete triplet ionization is observed in the inset of the Kr^{3+} spectrum, where it is just sufficiently separated from that of the 3d electrons. The spectrum for double O-shell ionization in Xe is shown in the inset of Fig. 4, together with the thresholds for formation of the $5s^25p^4$, $5s^15p^5$, and $5s^05p^6$ states. A few values obtained by Cairns et al. [18] in a photo-ionization experiment are also inserted in the figure. Their results are in excellent agreement with ours. However, the main conclusions from our coincidence measurements of the small angle inelastically scattered electrons in Kr and Xe and the ions formed are that we were able to demonstrate the presence of a minimum followed by a maximum in the contribution of the 4p–εd transitions in Kr and 5p–εd transitions in Xe. These minima and maxima were obscured in the photo-absorption measurements [20] by the rapidly rising contributions of 3d and 4d transitions in Kr and Xe, respectively. Furthermore, the results showed the existence of strong direct interaction between electrons in the outer and the inner shells, as opposed to a "shake off"-type interaction in Ar [15]. This gives evidence of the importance of the correlation between these shells of Kr and Xe, which is not considered in most of the calculations and is at least partially responsible for the discrepancies that exist between the experimental results of the oscillator strengths and those predicted by theory [21] and [22]. The electron–ion coincidence technique was also applied to study the K shell excitation of nitrogen and carbon monoxide by electron impact [23]. The study of the ionization of N_2 and CO by 10 keV electrons as a function of the energy loss was done by El-Sherbini and van der Wiel for the valence electrons [24] as well as for inner-shell electrons [25].

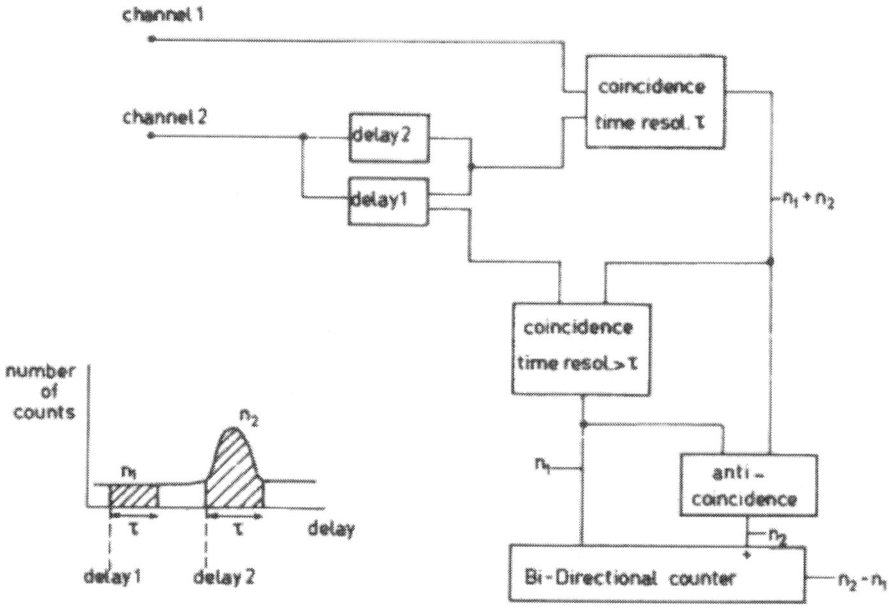

Figure 2. Block diagram of the coincidence circuit. Signal from the ion detector (channel 1). Signal from the electron detector (channel 2).

Our results on electron–atom ionization were the first of its type and corresponded well with those of photo-ionization by real and big synchrotron devices, but our apparatus was much faster and easier to operate. Our device was a sort of model synchrotron and in fact was considered to be **the first table-top synchrotron**.

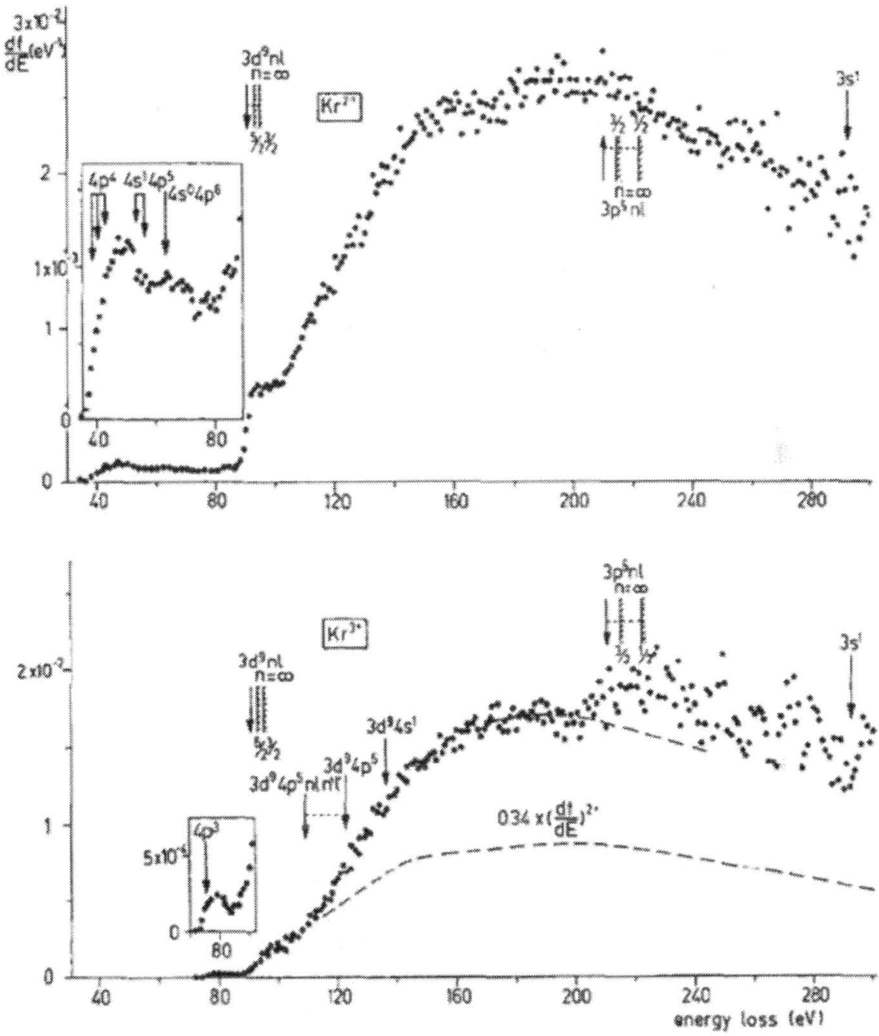

Figure 3. Oscillator-strength spectra of Kr^{2+} and Kr^{3+}. The inset of the upper figure shows the direct ejection of two N electrons below the $3d^9$ threshold in the Kr^{2+} spectrum. The inset of the lower figure shows the threshold for discrete triple ionization in the Kr^{3+} spectrum.

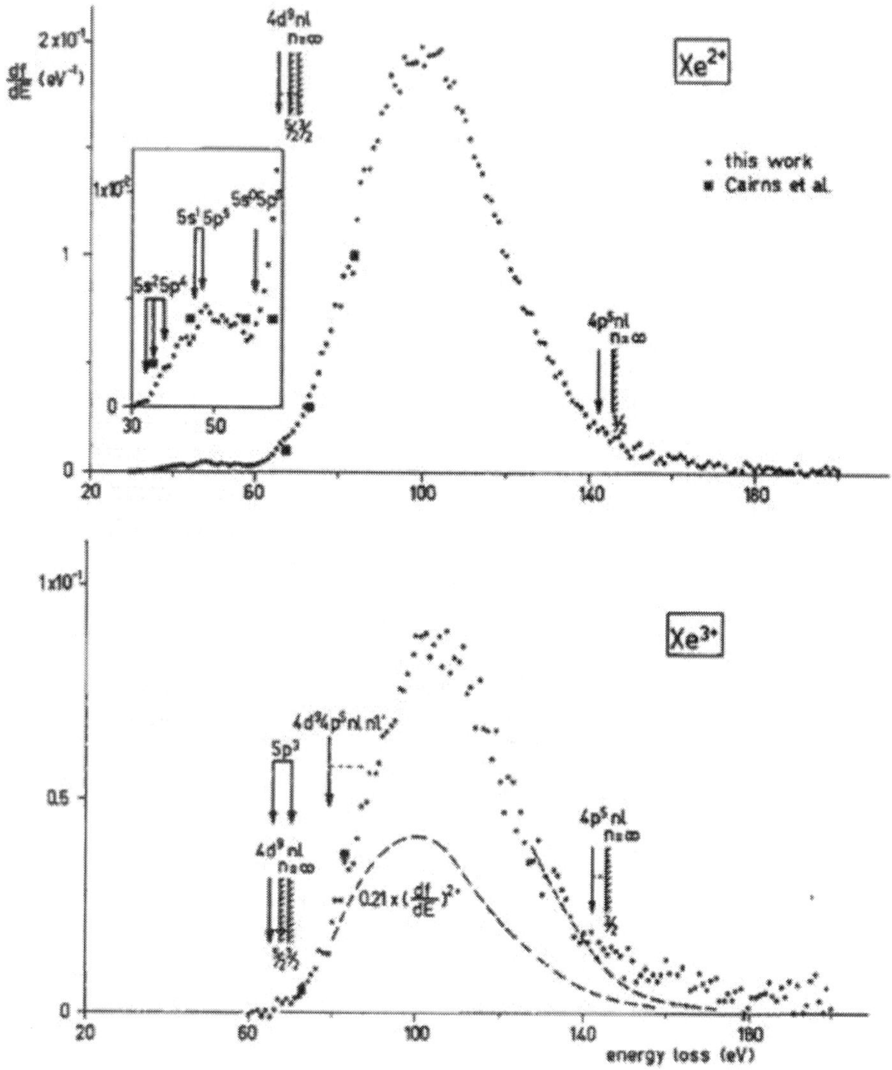

Figure 4. Oscillator-strength spectra of Xe^{2+} and Xe^{3+}. The spectrum for double O-shell ionization is shown in the inset of the figure together with the thresholds for formation of the $5s^25p^4$, $5s^15p^5$ and $5s^05p^6$ states. Our data are plotted together with a few values obtained by Cairns et al. [18], from a photo-ionization experiment.

ION–ATOM COLLISIONS

Collision processes between fast heavy atoms and ions can be simply described by the interactions between relatively fast protons and alpha particles with neutral atoms. Besides the normal excitations and ionizations which are analogous to what happens in electron–atom collisions, an extra phenomenon occurs, named charge exchange. The best way to describe both types of phenomena is in treating the three particles involved, viz the point charge projectile, the target atom, and the electron with one Hamiltonian. It is one closed system in which kinetic energy of the projectile is transferred into electronic excitation energy. The impact parameter treatment has proven very useful, see Bates [26]. It gave a semiclassical description of the collision process, with the external motions classically and the internal motions quantum mechanically. Due to the heavy mass of the proton or alpha particle, the kinetic energy of the projectile is much bigger than the electronic excitations concerned. Therefore, the trajectory of the projectile is considered rectilinear during the whole collision event. The projectile keeps constant velocity, approximately. The impact parameter ρ is defined as the distance between the trajectory and the target nucleus. The cross-section σ for transition of the electronic system from state i to state f is given by

$$\sigma_{if}(E) = 2\pi \int_0^\infty \rho P(\rho) d\rho \tag{5}$$

where E is the kinetic energy of the projectile in the center of mass system, and

$$P(\rho) = |a_{if}(\rho, t=\infty)|^2 \tag{6}$$

with

$$i\frac{d}{dt} a_{if}(\vec{R}, t) = \sum_k a_{ik}(\vec{R}, t) V_{fk}(\vec{R}) \exp(-i\Delta E_{kf} t) \tag{7}$$

$$\Delta E_{kf} = E_k - E_f \tag{8}$$

$\vec{R}$ is the distance between both nuclei; $V_{fk}(\vec{R})$ is the matrix element of the potential field of target particle scaled by $\frac{2m}{\hbar^2}$ between the target eigen states f and k; E_k and E_f are eigen energies of target particle; and a_{ik} is the amplitude of the target eigen functions. For kinetic energies E far above the threshold, we can apply the Dirac condition, which assumes that the most dominant transition is from the initial to the final state i.e.

$$a_{ik} = \delta_{kf} a_{ik} \tag{9}$$

This leads to the integral equation

$$ia_{if}(\vec{R}, t) = \int_{-\infty}^{t} a_{if}(\vec{R}, \tau) V_{if}(\vec{R}) \exp(-i\Delta E_{if}\tau) d\tau \tag{10}$$

In the first order Born approximation, we obtain

$$ia_{if}(\rho, t = \infty) = \int_{-\infty}^{+\infty} V_{fi}(\vec{R}) \exp(-i\Delta E_{if} t) dt \tag{11}$$

see Merzbacher [27]. Replacing t by $\frac{z}{u}$, where u is the velocity, one gets

$$ia_{if}(\rho, t = \infty) = \frac{1}{u} \int_{-\infty}^{+\infty} V_{fi}(\vec{R}) \exp\left(-i\Delta E_{if} \frac{z}{u}\right) dz \tag{12}$$

From this relation, the dependence of $P(\rho)$ on u can be deduced. Therefore, it will depend on

$$\frac{a\Delta E_{if}}{u} \text{ (in atomic units)} \tag{13}$$

One measures the effective interaction length "a" along the trajectory z, if the projectile passes by the target particle. This is the Massey Criterion. For large values of u, we see

$$\frac{a\Delta E_{if}}{hu} < 2\pi \tag{14}$$

which means that

$$P(\rho) \sim |a_{if}|2 \sim \frac{1}{u^2} \sim \frac{1}{E} \tag{15}$$

Decreasing speed coming from large values of u, one expects a maximum in $P(\rho)$ if $(a\Delta E_{if}/u \approx 2\pi)$, following the oscillatory behavior of $\exp(-i\Delta E_{if}z/u)$ as a function of u. This type of behavior has been studied by Hasted [28] and [29] who measured total cross-sections for exchange between various kinds of ions and neutral targets. Differential cross-sections, not only velocity dependent but also as a function of the scattering angle, have been measured by Morgan and Everhart [30] and by Kessel and Everhart [31].

Advances in this field were made by measuring electron capture by multiply charged ions. It attracted attention of many physicists in various fields of physics such as astrophysics, plasma physics, controlled thermonuclear fusion research, and X-ray laser production. When multiply charged ions collide with neutral particles (at low to intermediate impact velocities $u \leqslant 1$ au), capture reactions populating excited states in the projectile are very probable, see, for instance, Niehaus and Ruf [32] and Winter et al.[33]. For single electron capture, these reactions may lead to population inversion and are of importance in several schemes for the production of XUV and soft X-ray lasers. However, in these collisions, non-radiative (i.e. auto-ionizing) processes can be important, and competition with radiative processes occurs. Measurements of these non-radiative processes by Winter et al. [34] showed that the corresponding total cross-sections for the production of slow electrons were large and strongly charge state dependent. These results were interpreted by them to be the result of capture ionization, i.e., an Auger ionization in the short-lived quasi-molecule.

Let X^{z+} is the multiply charged ion and Y is the target atom, then the reactions can be followed by radiative emission

$$X^{z+}+Y \rightarrow X^{(z-1)+*}+Y^{+} \rightarrow X^{(z-1)+}+Y^{+}+h\nu$$

or by electron emission through one of the following channels

$$X^{z+}+Y \rightarrow X^{(z-1)+}+Y^{2+}+e \quad \text{(a)}$$

Auger ionization of the quasi-molecule formed during collision,

$$X^{z+}+Y \rightarrow X^{(z-1)+*}+Y+ \rightarrow X^{(z-1)+}+Y^{2}++e \quad \text{(b)}$$

Penning ionization after single electron capture,

$$X^{z+}+Y \rightarrow X^{(z-2)+**}+Y^{2+} \rightarrow X^{(z-1)+}+Y^{2+}+e \quad \text{(c)}$$

double electron capture into autoionizing states of the projectile,

$$X^{z+}+Y \rightarrow X^{(z-1)+*}+Y^{+} \rightarrow X^{(z-2)+**}+Y^{2+} \quad \text{(d)}$$

$$\rightarrow X^{(z-1)+}+Y^{2+}+e \quad \text{(e)}$$

electron capture followed by electron promotion [35] into auto-ionizing states of the projectile.

The measurements of Winter et al. [34] yielded only total cross-sections for Ne^{z+} ($z = 1$–4) and Ar^{z+} ($z = 1$–8) colliding at energies 100 keV and 200 keV, respectively, with noble gas atoms. However, data on the energy spectrum of the electrons are still needed to investigate these phenomena in more detail. Woerlee et al. [36] have extended the work by measuring energy spectra of electrons produced in collisions of multiply charged neon ions with noble gas atoms. Fig. 5 shows the experimental results for 100 keV Ne^{1-4+} on Ar. The spectrum consists of a continuous background on which peaks are superimposed. The spectra for Ne^{1+} and Ne^{2+} are almost identical, but large changes are seen when the projectile charge state is increased from 2+ to 3+ and 3+ to 4+. The largest changes are an increase in the continuum below ±20 eV, and an increasing number of peaks superimposed on the continua. The increase in the continuum below 20 eV is the result of capture ionization in the short-lived quasi-molecule [37]. The bars inFig. 5 indicate the positions of calculated transition energies corrected for a Doppler shift of −2.7 eV. The peaks observed in 100 keV $Ne^{3+,4+}$ on Ar shift to lower energies when the projectile energy is increased. This shift is equal to the kinematical shift, which would be expected, when the corresponding electrons are emitted by the projectile. Therefore, we concluded that the peaks originate from auto-ionizing states in the projectile, which decay after the collision has taken place. Since no photoabsorption data exist on the auto-ionizing states of multiply charged neon ions, we tried to calculate energy levels of doubly excited neon ions with a single configuration HF method. In order to determine the energies of the various levels, we included the electrostatic energy splitting due to the core electrons, see El-Sherbini and Farrag [38]. The energy splitting caused by the excited electrons is small and was not taken into account. We found that for Ne^{4+}–Ar, the peaks occur in the region for the calculated peak energies of Ne^{1+**}, Ne^{2+**}, and Ne^{3+**}, but Ne^{2+**} seems to cover most of the data. For Ne^{3+}–Ar, calculated energies of Ne^{1+**} and Ne^{2+**} appear in the region of the observed peaks.

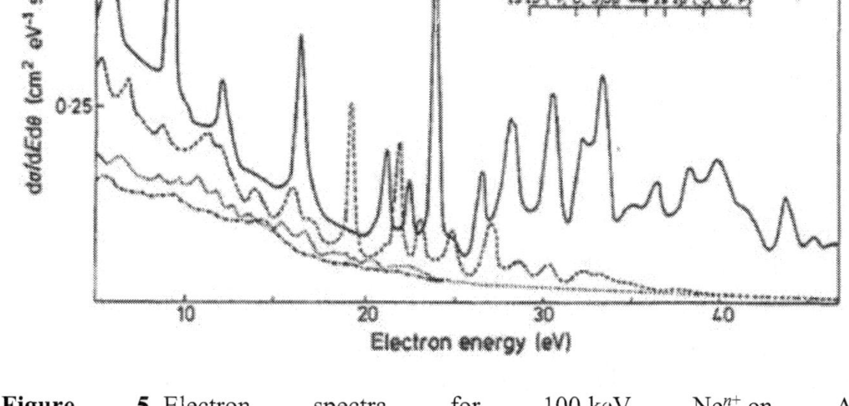

Figure 5. Electron spectra for 100 keV Ne^{n+} on Ar ($\vartheta = 90°$), - - - - $n = 1$; ········ $n = 2$; — — — $n = 3$; ——— $n = 4$. The bars in the figure indicate the positions of calculated transition energies corrected for a Doppler shift of −2.7 eV.

Further developments in this field were done by El-Sherbini et al. [39], where they measured target dependence of excitation resulting from electron capture in collisions of 200 keV Ar^{6+} ions with noble gases. The study shows strongly rising total capture excitation cross-sections and shifts in the post-collision projectile excited-state distributions to higher n levels with the increase in the target atomic number. Energy dependence of excitation and ionization resulting from electron capture in Ar^{6+}–H_2collision in the range of ion projectile energies 200–1200 keV was measured by El-Sherbini et al. [40]. These studies indicate that single electron charge transfer into excited states of the product ion is the most important inelastic process. Photon emission between 20 and 250 nm and slow electron and ion production cross-sections have been measured. The capture occurred mainly into $n = 4$ levels with the excitation of the higher angular momentum states dominating over most of the projectile energy range. The capture ionization cross-section is appreciable, amounting to 30–40% of the total excitation cross-section. These results are extremely valuable for the developments of controlled thermonuclear fusion reactors (see El-Sherbini [41]). To obtain more information about the coupling mechanisms, which gives rise to capture into excited states in ion–atom collisions at intermediate energies ($u \sim 0.5$ au), El-Sherbini and de Heer [42] measured photon emission in the spectral region between 60 and 100 nm in the collision of Ar^{q+} ($q = 1$, 2, and 3) with He and Ne at impact energies between 15 and 400 keV. The experimental results were explained qualitatively by considering the MO correlation diagram (Fig. 6). The emission cross-section for the collision of Ar^{q+} with He is shown in Fig. 7. It was often found that the cross-section for excitation decreases with the increase in the number of intermediate transitions required in order to reach the excited state. When there is a mechanism involving radial coupling leading from initial to final states, then it was found that the measured emission cross-section decreases with energy, where as mechanisms involving rotational coupling lead to cross-sections that increase with increasing energy up to 200 keV or more. The results have been of particular importance in evaluating theoretical models and have provided a valuable check of the range of validity of existing theories.

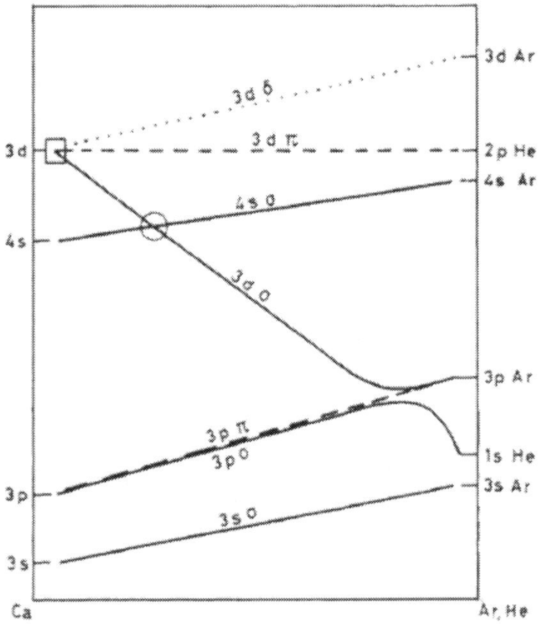

Figure 6. Diabatic MO correlation diagram for Ar–He system. The radial coupling occurs at the 3dσ–4sσ crossing and the rotational coupling occurs at the 3dσ–3dπ–3dδ crossing.

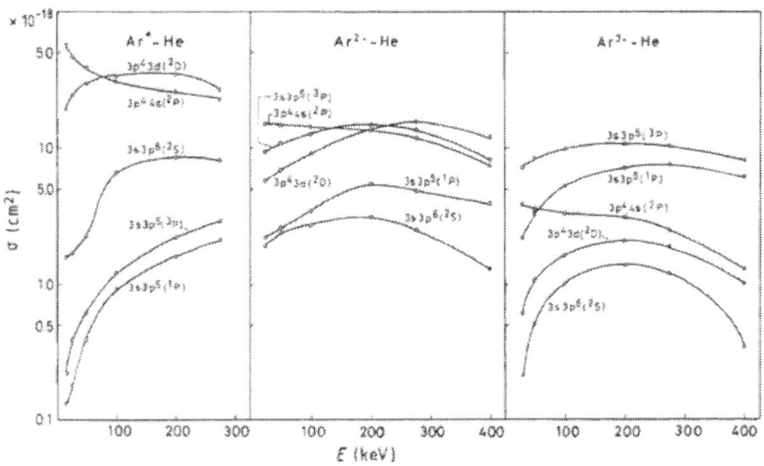

Figure 7. Emission cross-section for Ar II (3p^{4}4s ^{2}P), Ar II (3p^{4}3d ^{2}D), Ar II (3s3p^6 ^{2}S), and Ar III (3s3p^5 ^{3}P,^{1}P) states plotted against projectile energy in Ar^{q+}–He collisions.

ATOMIC STRUCTURE CALCULATIONS AND X-RAY LASERS

In the field of atomic collisions, as we noticed in the previous sections, much attention was paid to the excitation of noble gas atoms. A systematic study of the excitation process requires the knowledge of accurate dipole transition probabilities for spontaneous emission between the various configurations of the ions. Laser physics and astrophysics are other branches, which have stimulated more accurate atomic line strengths and transition probabilities calculations. Garstang [43] and [44] performed the first intermediate coupling calculations for Ne II. On this basis, Wiese et al. [45] composed their data compilations. However, the previously tabulated line strengths were in need of revision. In his work, Luyken [46] and [47] performed new calculations of line strengths and transition probabilities for Ne II and Ar II where specific configuration interactions were investigated and some effective operators were included. The results showed that the agreement with the experimental data was improved as compared with the earlier calculations. El-Sherbini [48], [49] and [50] has extended the work of Luyken to the calculation of transition probabilities and radiative lifetimes for Kr II and Xe II. He used "exact" intermediate coupling wave functions to describe the various states [48] :

$$\Psi(J,M) = \sum \alpha_i |p^4 L_i^c S_i^c; \quad l_r \frac{1}{2}; \quad L_i S_i JM > \tag{16}$$

where α_i is the expansion coefficient, J is the total angular momentum, M is the magnetic quantum number, L_i^c and S_i^c are the total orbital and spin angular momentum of the core electrons, l_r is the orbital angular momentum of the running electron, and $L_i$and S_i are the orbital and spin angular momentum of the pure L–S bases states on which the "exact" $\Psi(J, M)$ is expanded. The transition probability between two states with summation indices i and j refer to the upper and lower level, respectively, is given by

$$A(J_u, J_l) = \frac{64\pi^2}{3h\lambda^3(2J_u + 1)} S(J_u, J_l) \tag{17}$$

where $S(J_u, J_l)$ is the line strength and J_u, J_l are the total angular momentum of the upper and lower states, respectively. The line strength is given by El-Sherbini [48]

$$S(J_u, J_l) = e^2 \left| \sum_{i,j} \alpha_i^* \alpha_j (-1)^{S_j + J_u + l_{ru} + L_j^c} \delta(S_i, S_j) \delta\left(L_i^c, L_j^c\right) \begin{Bmatrix} S_j L_j J_l \\ 1 J_u L_i \end{Bmatrix} \right.$$
$$\begin{Bmatrix} L_j^c l_{rl} L_j \\ 1 L_i l_{ru} \end{Bmatrix} \begin{pmatrix} l_{ru} 1 l_{rl} \\ 000 \end{pmatrix} [(2J_u + 1)(2J_l + 1)(2L_i + 1)$$
$$\left.(2L_j + 1)(2l_{ru} + 1)(2l_{rl} + 1)]^{1/2} \right|^2 \left(\int_0^\infty R_{l_{ru}}(r) r R_{l_{rl}}(r) dr \right)^2 \tag{18}$$

where J_u, J_l and l_{ru}, l_{rl} are, respectively, the total angular momentum of the states and the orbital angular momentum of the running electron in the upper and lower states. $R_l ru$(r) and $R_l rl$(r) are the one electron radial wavefunctions in the two different states.

The lifetime τ_u of the upper state is given by El-Sherbini [49]

$$\tau_u = \frac{1}{\sum_l A(J_u, J_l)} \tag{19}$$

The parametric potential method was used to calculate the radial part of the wave function [51], while the method of least squares fit of energy levels [52] was applied in obtaining the angular part of the wave function. The results obtained in intermediate coupling showed a much better agreement with the experimental data than those using pure LS-coupling wave functions. Further improvements in the atomic structure calculations of Kr II were obtained by El-Sherbini and Farrag [38] when including configuration interaction effects. The results showed that the $4s^2 4p^4(^1D)4d\ ^2S_{1/2}$ level is strongly perturbed through interaction with the $4s4p^6\ ^2S_{1/2}$ level, in agreement with the earlier predictions from the Kr II analysis. Theoretical investigations of the $5s^2 5p^4 5d + 5s^2 5p^4 6s + 5s5p^{6+}$ level structure in Xe II were performed by El-Sherbini and Zaki [53]. Taking into account, configuration-interaction effects in the calculations showed that some observed energy levels of the $5p^4 5d$ configuration were not correctly designated. A strong interaction between the $5p^4 5d$ and $5s5p^6$ configurations was also reported. Moreover, the calculated energies of the 6s and 5d levels were improved considerably by introducing configuration interactions into the calculations. The presence of strong configuration interaction between the $4s4p^6$, $4p^4 4d$, and $4p^4 5s$ configurations in singly ionized krypton [38] makes it difficult to perform accurate calculations for the energies, pumping rates, and lifetimes of levels in these configurations. Therefore, it was important to improve upon the previous calculations, see El-Sherbini [54] and [55], on the low lying $4p^4 4d$ and $4p^4 5s$ laser levels in this ion. Therefore, multi-configuration Hartree–Fock (MCHF) calculation in order to determine the lifetimes of these laser levels was done by El-Sherbini [56]. The results show that some of these levels are metastable. They also suggest a two-step excitation from the ground state of the ion to the $4p^4 5p$ level involving some intermediate metastable states as a possible laser excitation mechanism.

Further developments in the field of atomic structure calculations were done by the studies of excitation of electrons in atomic isoelectronic sequences [57], [58] and [59]. These studies are essential not only for better understanding of atomic structure and ionizing phenomena, but also they provide new laser lines which could be extended into the X-ray spectral region [60] and [61]. This in turn will help in the development of X-ray laser devices. Once X-ray lasers become reliable, efficient, and economical, they will have several important applications. First and foremost, their short wave lengths, coherence, and extreme brightness should allow the exploration of living structures much smaller than one can see with optical methods. They will also have important applications in high resolution atomic spectroscopy, diagnostics of high density plasmas, radiation chemistry, photolithography, metallurgy, crystallography, medical radiology, and holographic imaging. Shortly after the demonstration of the first soft X-ray amplification in neon-isoelectronic selenium by Mathews et al. [62], extensive work was done both theoretically and experimentally on other systems [63] and [64]. Progress toward the development of soft X-ray lasers with several plasma-ion media of different isoelectronic sequences was achieved at many laboratories [65] and [66]. A soft X-ray laser transitions in the Be-isoelectronic sequence were proposed by Krishnan and Trebes [67]. They suggested that intense line radiation from plasmas of Mn VI, P IV, Al V, Al V III, Al IX, and Al XI may be used to selectively pump population inversions in plasmas of Be-like C III, N IV, F VI, and Ne VII and Na VIII. Lasing in the soft X-ray region is then possible on 4p–3d and 4f–3d (singlet and triplet) transitions. Short wave length laser calculations in the beryllium sequence were done by Feldman et al. [68]. They calculated gain at a number of different temperatures and electron densities for the 3p–3s laser transition in the highly charged ions of Be-sequence. Al-Rabban [69] has extended both the work of Krishnan and Trebes [67] and Feldman et al. [68], to the higher members of the Be-isoelectronic sequence and to more transition states (which are promising for X-ray laser emission). She carried out an ab initio multi-configuration Hartree–Fock calculations of energy levels, atomic oscillator strengths, and radiative lifetimes for singly and doubly excited states in Be I and Be-like ions. Configuration interaction effects between the various configurations were included using the computer program code CIV3 described by Hibbert [70]. In this code, the N-electron energies and eigenfunctions are obtained by diagonalizing the Hamiltonian matrix, which may have quite large dimensions. The choice for the spatial (radial) part of the single particle wave functions is based on expansions in Slater-type orbitals [71]:

$$P_{nl}(r) = \sum_{j=1}^{k} C_{jnl} r^{I_{jnl}} \exp(-\xi_{jnl} r) \tag{20}$$

The coefficients in the expansion C_{jnl}, I_{jnl} as well as ξ_{jnl} in the exponents are treated as variational parameters.

Investigations of the possibilities of obtaining population inversion and laser emission could be achieved by calculating the level population of the excited states. These calculations were done by the group of atomic physics at the Physics Department of the Faculty of Science – Cairo University, solving the coupled rate equations [72]

$$N_j\left[\sum_{i(j} A_{ji} + N_e\left(\sum_{i(j} C^d_{ji} + \sum_{i)j} C^e_{ji}\right)\right] = N_e\left(\sum_{i(j} N_i C^e_{ij} + \sum_{i)j} N_i C^d_{ij}\right) + \sum_{i)j} N_i A_{ij} \quad (21)$$

where N_j is the population density of level j, A_{ji} is the spontaneous decay rate from level j to level i , C^e_{ji} is the electron collisional excitation rate coefficient, C^d_{ji} is the electron collisional de-excitation rate coefficient, and N_e is the plasma electron density. The gain coefficient (α) for Doppler broadening of the various transitions is given by Elton [73]:

$$\alpha = \frac{\lambda^2_{lu}}{8\pi}\left(\frac{M}{2\pi k T_i}\right)^{1/2} A_{ul} N_u F \quad (22)$$

where M is the ion mass, λlu is the transition wave length in cm, T_i is the ion temperature in K, u and l represent the upper and lower transition levels, respectively,N_u is the population of the upper level, and F is the gain factor.

Vriens and Smeets [74] gave empirical formulas for the calculation of rate coefficient in hydrogen atom. Their work was extended by Allam [75] to be valid for atoms with one electron outside a closed shell and also for two-electron atoms (ions). Allam [75] adopted the method of Palumb and Elton [76] for modeling plasmas of helium-like and carbon-like ions, and he has developed a computer program (CRMOC) in order to calculate excitation and de-excitation rate coefficients for two-electron system. In his program which was developed for collisional radiative model calculations, the principal quantum numbers of the excited states were replaced by effective quantum numbers. Using the above theoretical schemes, the atomic physics group was able to extensively investigate the possibility of X-ray laser emission in several isoelectronic systems, see for exampleFig. 8 and Fig. 9. The studies include helium isoelectronic sequence [77], beryllium isoelectronic sequence [69] and [78], boron isoelectronic sequence [79], [80] and [81], carbon isoelectronic sequence [82], sodium isoelectronic sequence [83], [84] and [85], magnesium isoelectronic sequence [86], [87] and [88], aluminum isoelectronic sequence[89], silicon isoelectronic sequence [90], [91] and [92], sulfur isoelectronic sequence [93], potassium isoelectronic sequence [94], scandium isoelectronic sequence [95], and nickel isoelectronic sequence [96]. Most of the heavy members of the isoelectronic sequences studied radiate in the XUV and Soft X-Ray spectral regions (λ between 50 and 1000 Å). The reported stimulated emission transitions in these ions indicate that some of the transitions are promising and could lead to progress toward the development of XUV and Soft X-Ray lasers.

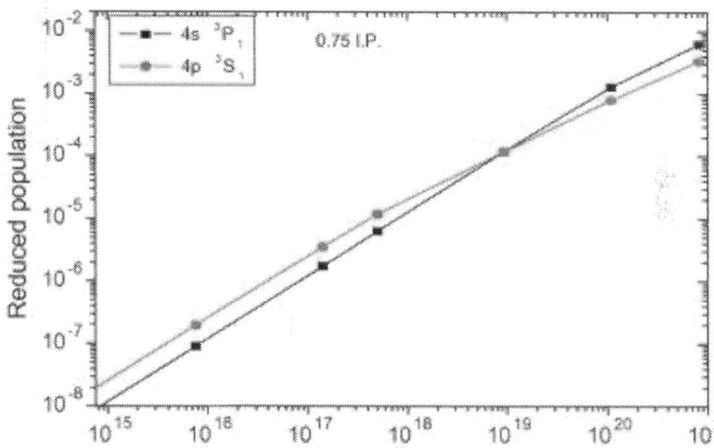

Figure 8. Reduced fractional population for selected levels of Ni^{14+} ions at electron temperature 3/4 the ionization potential.

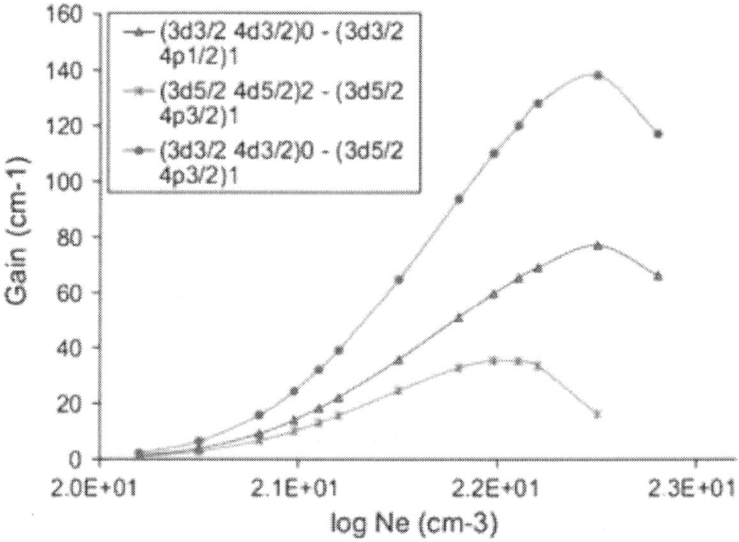

Figure 9. Gain coefficient of laser transitions against electron density at temperature 2 keV in E_{u}^{35+} ions.

LASER-INDUCED BREAKDOWN SPECTROSCOPY (LIBS)

Laser-induced breakdown spectroscopy is a form of optical (atomic) emission spectroscopy [97]. It is a technique based on utilizing light emitted from plasma generated via interaction of a high power lasers with matter (solids, liquids or gases). Assuming that light emitted is sufficiently influenced by the characteristic parameters of the plasma, the atomic spectroscopic analysis of this light shows considerable information about the elemental structure and the basic physical processes in plasmas. There is a growing interest in LIBS, particularly in the last 20 years because of its applications in the laboratory and in industry, art, environment, medicine, and forensic sciences [98], [99] and [100]. Most commonly, LIBS has been applied to sensitive elemental analysis of solids, conductors and non-conductors, as well as liquid and gaseous samples [101]. It has many practical advantages over more conventional elemental analysis techniques. LIBS has been utilized to analyze thin metal films [102], and it has found more and more applications in monitoring of industrial processes [98], characterization of jewellery products [103], soil studies [104], pulsed laser thin film deposition [105], quality control of pharmaceutical products [100], cleaning [106], and in situ planetary exploration [107].

An enhancement of the LIBS sensitivity was achieved by introducing the double pulse technique [108]. The double pulse (DP)-LIBS configuration, which makes use of two laser pulses separated by a suitable temporal delay instead of a single pulse for inducing the plasma, was reported to give a substantial enhancement of the signal to noise ratio with respect to single pulse (SP)-LIBS configuration with a corresponding improvement of the limits of detections [109]. The double pulse laser ablation (DPLA) approach in relation to the spectral analysis was first reported by Piepmeier and Malmstadt [110]. However, the first systematic investigation of (DPLA) was reported by Sattmann et al.[111]. They performed a quantitative microchemical analysis of low-alloy steel with single and double laser pulses, where they found that the analytical performance was considerably improved by the double pulse technique. The great contribution to the development of (DPLA) for practical analysis was made by Petukh et al. [112]. They compared radiation of plasma flares produced on exposure of metals to laser radiation in a monopulse generation mode in the case of single and double pulses with change in air pressure. They observed in the case of double pulses increases in the duration and the intensity of the radiation of the spectral lines. For elucidation of the double pulse laser ablation (DPLA) mechanisms, see, for instance, St-Onge et al. [113] and Noll [114]. DP-LIBS technique was also applied for the fabrication of nanosize particles. Tarasenko et al. [115] studied and analyzed the capabilities of laser ablation in liquids for fabrication metallic and composite nanoparticles. The technique offers the better controle over the particle formation process. They found that the mean size of the nanoparticles and their stability could be controlled by proper selection of the parameters of laser ablation such as temporal delays between pulses, laser fluence, and the sort of liquid used. Therefore, the optimal conditions favoring the formation of nanoparticles with a desired structure could be reached.

Parallel to the work on atomic structure calculations by our atomic physics group at the physics department, the group was also involved in the study of the physical parameters of plasmas generated by high power laser irradiation of solid targets (plasma diagnostics), applying the (LIBS) technique. The spectroscopic plasma diagnostics which is essentially based on the measurements of the optical radiation emitted from the plasma enables the group to obtain simultaneously a large amount of information about the plasma without disturbing it. Spectral fingerprints of optical plasma emission provide information about the physical and chemical processes that occur in the plasma. The spectra can contain individual spectral lines, band, or continuum radiation. Plasma emits line radiations resulting from bound–bound electronic transitions and continuum radiations resulting from free-bound and free–free electronic transitions. However, utility of spectroscopic diagnostics depends upon the knowledge about radiative behavior of atomic and molecular species and type of equilibrium attained in the plasma. It is assumed that the plasma in our laboratory (laboratory of lasers and new materials at the physics department) is in local thermodynamic equilibrium (LTE). In local thermodynamic equilibrium, all the species in the plasma, i.e., electrons, ions, and neutrals are in thermodynamic equilibrium except the radiation. This condition generally is observed to be valid in a collision dominated plasma such as high-pressure arc plasma produced in plasma torches. Small size of such plasmas allows radiation to escape to the surroundings. In (LTE) plasmas, the number of electronic transitions due to collisions between the first excited states and the fundamental level is 10 times larger than the number of transitions due to spontaneous emission. Collisions are mainly responsible for excitation and de-excitation, ionization, and recombination. The two main parameters that characterized the state of the plasma are namely the plasma temperature and the electron density. Knowledge of the temperature leads to understand the plasma processes occurring such as vaporization, dissociation, ionization, and excitation. The optical emission spectroscopic (OES) method for the determination of the plasma temperature is based on the measurements of the intensity of the spectral lines. In optically thin plasma, the integrated intensity of an atomic emission line is related to excitation energy, population density of upper state and transition probability as given by

$$I_{ul} = \frac{1}{4\pi} A_{ul} n_u h\upsilon_{ul} L \ (\mathrm{W/m^2\text{-}ster}) \tag{23}$$

where I_{ul} is the line intensity of transition from upper level u to lower level l integrated over the plasma length L, A_{ul} is the spontaneous transition probability, n_u is the density of atom excited in the upper energy level u, and $h\upsilon_{ul}$ is the energy of each emitted quantum. The measurement of I_{ul} gives only the population of upper level u. When the thermal plasma is in (LTE), the density of atoms excited to the upper level is given by the Boltzmann distribution function:

$$n_u = \left(\frac{n_0}{Z_0}\right) g_u \exp\left(\frac{-E_u}{kT}\right) \tag{24}$$

where n_0 is the total density of atoms, g_u is the statistical weight of the upper state, $E_u$is the energy of upper state, k is Boltzmann constant, and Z_0 is the partition function defined by

$$Z_0 = \sum_u g_u \exp\left(\frac{-E_u}{kT}\right) \tag{25}$$

Substituting the value of n_u into Eq. (23), we get

$$I_{ul} = \frac{1}{4\pi} \frac{hc A_{ul}}{\lambda_{ul}} \frac{n_0 L}{Z_0} g_u \exp\left(\frac{-E_u}{kT}\right) \tag{26}$$

In case of the evaluation of absolute line intensity, one should know the initial composition, pressure and wave length of the emission line. The values of A_{ul}, g_{ul}, andE_u can be obtained from spectroscopic tables. However, one must also know the plasma length, and an absolute spectral radiance calibration must be performed using a standard source. For relative line intensities measurement of the same species and stage of ionization, one needs not to know the values of partition function, n_0, and plasma emitting length. The ratio of two emission lines I_1 and I_2 is given by

$$\frac{I_1}{I_2} = \frac{g_1 A_1 \lambda_2}{g_2 A_2 \lambda_1} \exp\left(\frac{E_2 - E_1}{kT}\right) \tag{27}$$

The terms in Eq. (26) can be arranged as

$$\frac{I_{ul}\lambda_{ul}}{g_u A_{ul}} = \frac{hcn_0 L}{4\pi Z_0} \exp\left(\frac{-E_u}{kT}\right) \tag{28}$$

and therefore, we can write

$$\ln\left(\frac{I_{ul}\lambda_{ul}}{g_u A_{ul}}\right) = B - \frac{E_u}{kT} \tag{29}$$

This is an equation of straight line where $B = \ln\,(hcn_0L/4\pi Z_0)$ is a constant. Ifln($I_{ul}\lambda_{ul}/g_uA_{ul}$) values are plotted against E_u, the temperature is given by the reciprocal of the slope of the straight line. This is called the atomic Boltzmann plot method. The other key parameter in the diagnostics of plasma is the electron density. Determination of the electron density n_{ek} is based on the broadening of emission lines from the plasma. It is assumed that the Stark effect is the dominant broadening mechanism, in comparison with Doppler broadening and the other pressure broadening mechanisms, due to collisions with neutral atoms. The validity of this assumption was generally admitted in works on (LIBS) and is justified in various studies [116] and [117]. For the linear Stark effect (hydrogen and hydrogenic ions), the following equation is valid [116]

$$n_e = C(T, n_e)\Delta\lambda_s \tag{30}$$

where $\Delta\lambda_s$ is the full width at half maximum (FWHM) of the spectral line and $C(T, n_e)$ is a coefficient that is only a weak function of electron density and temperature. For elements other than hydrogen and hydrogen-like ions, the quadratic Stark effect acts on the total half width at half maximum (HWHM), due to collisions with electrons and ions, ω_{total} is approximately given by Griem [117]

$$\omega_{total} = [1 + 1.75A(1 - 0.75R)]\omega_s n_e / n_e^{ref} \tag{31}$$

where ω_{total} is the electron impact (half) width, A is the ion broadening parameter, ω_s is the Stark width, n_e^{ref} is a reference electron density, usually of the order of 10^{16} or 10^{17} cm^{-3}, and R is the ratio of the mean distance between ion and Debye radius. Eq.(31) is used in LIBS to determine the electron density from experimentally measured line widths of selected lines. For accurate measurements of the electron density, spectral lines as isolated as possible and emitted in optically thin conditions have to be selected. Stark broadening of isolated spectral lines of non-hydrogenic neutral atoms and ions is due mainly to electrons. As a consequence, the contribution of quasi-static ions was generally neglected, and hence, the Lorentzian FWHM can be approximated by

$$\omega_{total} = \omega_s (n_e / n_e^{ref}) \tag{32}$$

Our group at the laboratory of lasers and new materials at the physics department has studied emission spectra from laser-induced titanium plasma [118] and measured population density and temperature of argon metastable ($^1S_3$) state using tunable diode laser-absorption diagnostic technique [119]. An elemental analysis of some minerals using laser-induced breakdown spectroscopy (LIBS) was performed at the laboratory by El-Sergany et al. [120].

Self-absorption effect can distort the spectral line shape and therefore produces apparently an increase in the line width and a decrease in the line intensity, Griem [116]. It is originated mostly from cooler boundary layer of the plasma which contains much lower population density [117]. This effect will mislead investigators and can give in accurate results for the plasma parameters. Therefore, for the work on LIBS, one has to check the presence of self-absorption. Our group has evaluated self-absorption coefficient of Aluminum emission lines in laser-induced breakdown spectroscopy (LIBS) measurements, see El-Sherbini et al. [121]. Hydrogen lines, especially those of the Balmer series exhibit linear Stark effect, are the most strongly broadened lines, and they are easy to measure. Therefore, they are oftenly used to determine the electron density. Our group, El-Sherbini et al. [122], has measured the electron density in a laser produced plasma experiment using the Stark broadening of H_α-line at 656.27 nm. This line is produced from the interaction of a Q-switched Nd-Yag laser beam at the fundamental wave length of 1.06 μm with a plane solid aluminum target in a humid air Fig. 10. In this experiment, light emitted from the plasma plume is collected by a lens and optical fiber arrangement using an imaging spectrograph with ICCD camera, see Fig. 10. The wave length scale was calibrated with a low pressure Hg-lamp. The emitted light was collected in the wave length region from 200 to 1000 nm. The gain of the camera was kept fixed at a maximum level of 250. The measurement was confirmed by observing the spectra emitted at wave length regions from plasma at delay times from 30 to 50 μs which gave the same band width. Identification of the different spectral lines was carried out using a software spectrum analyzer (version 1.6). The agreement between the measured electron density from both the H_α-line and the Al II-line at 281.62 nm confirms the reliability of utilizing the H_α-line as an electron density standard reference line in LIBS experiments.

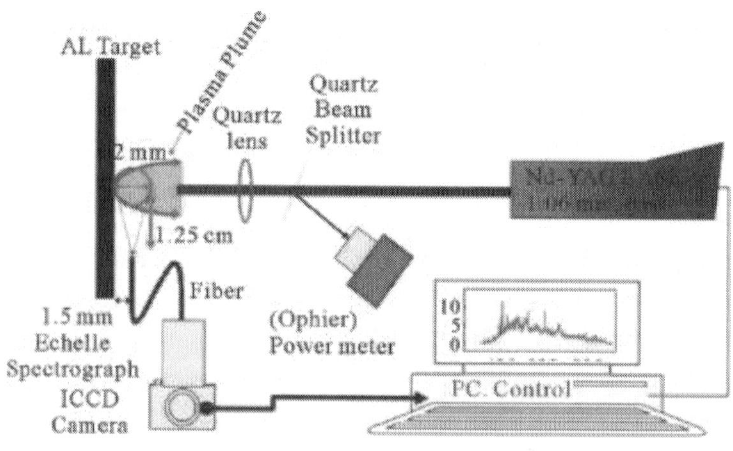

Figure 10. Experimental set up.

Our group has also measured the Stark broadening of atomic emission lines in non-optically thin plasmas by of laser-induced breakdown spectroscopy, El-Sherbini et al.[123]. An assessment of LIBS diagnostics of plasma using the hydrogen H_α-line at different laser energies in air was performed by the group, El-Sherbini et al. [124]. Moreover, the atomic physics group has applied the diode laser atomic absorption spectroscopy (DLAAS) technique to assess the degree of optical opacity of plasma at the wave length of H_{α}-line, El-Sherbini et al. [125]. They found that the plasma associated with metallic targets is almost optically thin at the H_{α}-line over all fluencies and at delay times $\geqslant 1$ μs, but rather thick for hydrogen-rich targets (plastic and wood) over all delay times and fluencies.

Recent measurements of plasma electron temperature utilizing magnesium lines appeared in laser produced aluminum plasmas were done by the group, El-Sherbini et al.[126]. This work shows that the Mg I and Mg II lines appeared at the short wave length region of the LIBS spectrum are good candidates for measuring the temperature of the plasma in LIBS experiments, but after correction against self-absorption. The spectrometric measurements of plasma parameters utilizing the target ambient gas O I and N I atomic lines show the reliability in the values of the electron density and the temperature of the plasma generated by the interaction of laser beam with solid targets, El-Sherbini et al. [127].

Advances in the LIBS measurements at our laboratory of lasers and new materials in the physics department at the faculty of science – Cairo University were achieved by studying the X-ray/particle emission from plasmas produced by laser irradiating nano-srtuctured targets, Hegazy et al. [128]. In this experiment, nano-copper structures evaporated onto copper bulk disks and nano-gold structures evaporated onto gold ones were used. An ion collector and X-ray semiconductor diode were used to study the ion and X-ray emission, respectively. A comparison of both studies in the case of nano-structured targets and bulk targets was performed at different laser fluencies (1×10^9–1×10^{12} W/cm^2) on the target. A 20% increase in the X-ray emission for nano-gold with respect to the bulk gold was observed; however, the X-ray emission in the nano-copper and copper was the same. At high laser intensities, the presence of non-linear processes in the preformed plasma may significantly increase the temperature of the fast electrons, and therefore mainly, the hard components of X-ray radiation at the fast ion emission are produced that does not happen in low energy nano-second experiments. Another progress was also recently achieved at our laboratory, by observing enhancements in LIBS signals from nano vs. bulk ZnO targets and nano-based targets, El-Sherbini et al. [129]. The study revealed that the signal enhancement cannot be attributed to the plasma temperature difference or the difference in the electron density. The signal enhancement depends only on the relative atomic concentration in the plasma created from the nano-based material with respect to the bulk-based plasma. This can be qualitatively explained in terms of the collisonal radiative modeling; for an electron density in the order of 10^{17} cm^{-3}, the collision processes are the dominant ones. Therefore, the enhanced emission of Zn I-lines from the nano-based target Fig. 11 could be attributed to the higher concentration of neutral atoms in the nano-based material plasma with respect to the corresponding bulk-based ZnO material.

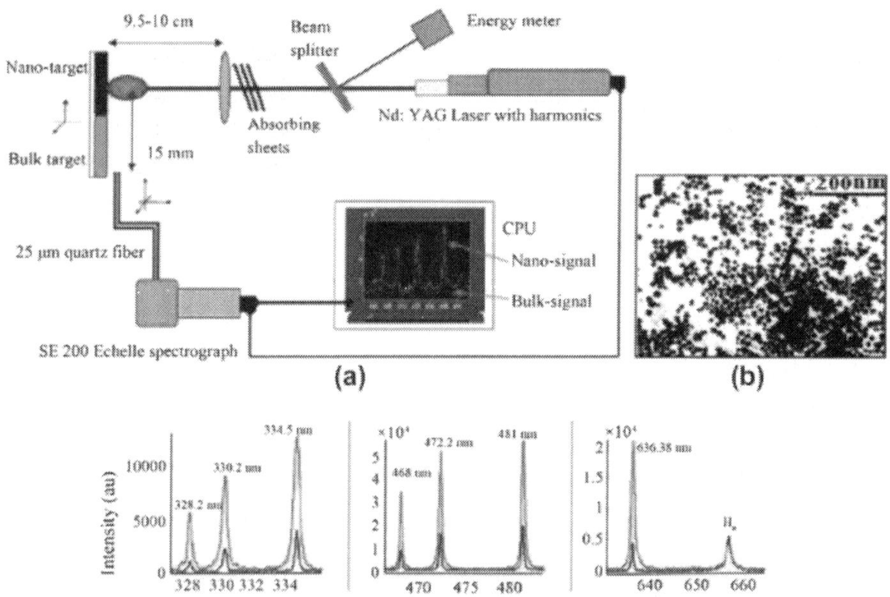

Figure 11. (a) experimental setup; (b) TEM image of the ZnO-20 nm size; (c) enhanced emission of the Zn I – line from the nano-based target (red color) in comparison to that from the bulk-based material (black color).

The evolution of Al plasma generated by Nd-YAG laser radiation at the fundamental wave length was studied at our laboratory by Hegazy et al. [130]. The results showed that the plasma temperature and the electron density are strongly dependent on the time and on the laser energy. More recently, another study was done at our laboratory by the same authors on titanium targets [131]. They investigated the spectral-evolution of nano-second laser interaction with Ti target in air. In this study, time resolved optical emission spectroscopy (OES) has been successfully employed to investigate the evolution of plasma produced by IR- and visible-pulsed laser beams irradiating a titanium target in ambient air at atmospheric pressure. The characterization of the plasma-assisted pulsed laser ablation of the titanium target was discussed. The obtained temperature was in a good agreement with the one obtained from Ti II spectral lines previously suggested by Herman et al. [132]. Moreover, the Stark broadening method has been employed in the experiment for electron density measurements.

In conclusion, plasma diagnostic technique with LIBS is essential for accurate determination of plasma parameters such as temperature and electron density. It can give more information about the atomic processes and the dynamic nature of the plasma. This information plays a major role in astrophysics and in controlled thermonuclear fusion research.

LASER COOLING AND BOSE–EINSTEIN CONDENSATION

One of the research highlights in atomic physics in recent years is the study of laser cooling and Bose–Einstein condensation. As early as 1917, Einstein [133] had predicted that momentum is transferred in the absorption and emission of light, but it was not until the mid eighties that such optical momentum transfer was used to cool and trap neutral atoms [134] and [135]. By properly tuning laser light close to atomic transitions, atomic samples can be cooled to extremely low temperatures [136]. In 1985, National Bureau of standards group [137] cooled atoms in a thermal atomic beam in the range of 50–100 m K by irradiating them with a beam of counter-propagating against their motion. If the laser is tuned to the low frequency side of an atomic resonance, an atom moving against the direction of a laser beam will see the beam Doppler-shifted into resonance, while the beam co-propagating with the atom will be Doppler-shifted out of resonance. Thus, the atom will preferentially scatter photons from the beam opposing the direction of motion. A step further was cooling in three dimensions, and this was accomplished by creating three sets of counter-propagating beams along the x, y, and z axes. The idea was suggested by Hansch and Schawlow [138] and was demonstrated by Chu et al. [139]. Because the cooling force is viscous (linearly proportional to the velocity of atoms for low velocities), the laser beams that generate the drag force was named "optical molasses."

A substantial improvement in laser cooling was achieved with optical molasses (OM), where three intersecting orthogonal pairs of oppositely directed laser beams are used to severely restrict the movement of atoms in the intersection region. With this technique, atoms could be cooled to temperatures between 1 and 0.1 μK. The first cooling and trapping of neutral atoms (sodium atoms) was accomplished with optical molasses in combination with magneto-static fields Fig. 12, and the technique was named magneto-optical trapping MOT [140].

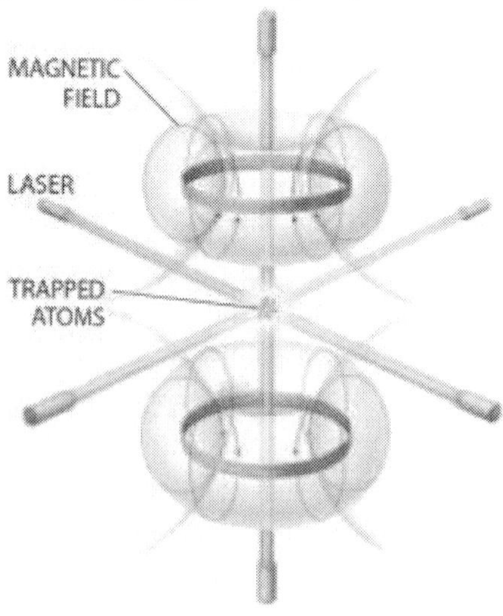

Figure 12. The technique of magneto-optical trapping MOT, Scientific American (January 1998).

When atoms in a gas are cooled to extremely low temperatures, they will (under the appropriate condition) condensate into a single quantum mechanical state known as a Bose–Einstein condensate (BEC). This phenomenon was predicted by Nath Bose and Albert Einstein in 1925 when they pointed out that at low temperatures particles in a gas could all reside in the same quantum state. The Bose–Einstein condensation exhibits a new state of matter which occurs at extremely low temperatures when the de Broglie wave length of atoms becomes comparable to the average distance between them Fig. 13. The temperatures reached by laser cooling are impressively low, but they are not low enough to produce BEC in gases at the densities that are realizable experimentally. In the experiments performed to obtain BEC of alkali gases, evaporative cooling which was first suggested by Hess [141] was used after laser cooling. With this additional cooling, the temperature reached about 2 nK, which was sufficiently low to form BEC. The basic physical effect in evaporative cooling is that if particles escaping from a system have energy higher than the average energy of particles in the system, the remaining particles are cooled. If one makes a hole in the magneto-optical trap (MOT), only atoms with an energy at least equal to the energy of the trap at the hole will be able to escape. In practice, one can make such a hole by applying radio-frequency radiation that flips the spin state of an atom from a low-field seeking one to a high field seeking one, thereby expelling the atom from the trap [142]. In order to diagnose the dense and cold samples of trapped atoms, the time of flight (TOF) method is used [143]. The (TOF) method was capable of mapping out the velocity distribution for both hyperfine and ground states of dilute gases of alkali atoms along their path through the trap, see Fig. 14. The peculiar BEC gaseous state was created and diagnosed in the laboratory for the first time in 1995, by Carl Wieman, Cornell, and co-workers [144] at the JILA laboratory in Boulder and by Ketterle and co-workers [145] at the MIT, using the powerful laser cooling method together with evaporative cooling method. Nowadays, the phenomenon of BEC has become an increasingly active area of research both experimentally and theoretically. Research work on this form of matter is relevant to many different areas of physics – from atomic clocks and quantum computing to superfluidity, superconductivity, atom lasers, and quantum phase transition.

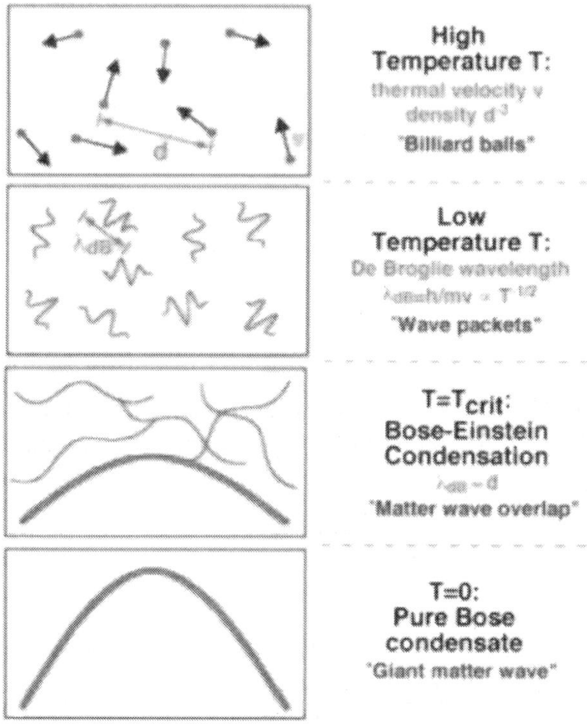

Figure 13. A diagram showing the properties of particles in a gas at various temperatures, Ketterle [142].

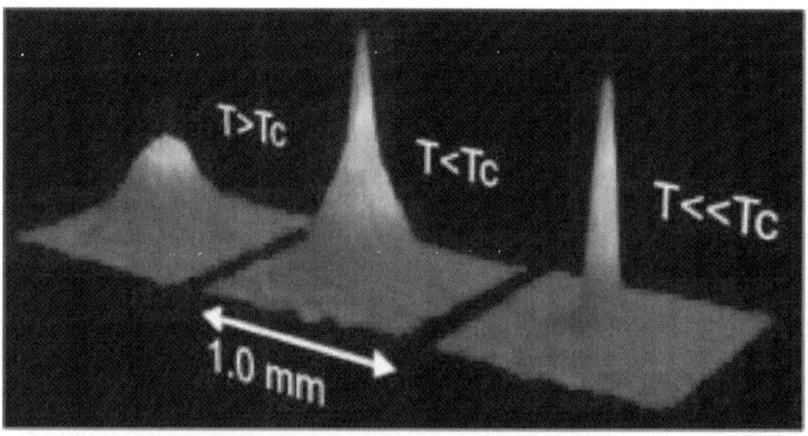

Figure 14. The density profile of the condensate after time of flight expansion, giving emphasis to the velocity distribution, Ref. [144].

The pioneering work on BEC by Wieman, Cornell and Ketterle and others [144], [145] and [146] was centered on atoms that were bosons, particles with integer spins. But atomic physicists have since extended their work to create Fermi gases from atoms that are fermions, particles with half odd integer spins. The advances came in 2003 by Deborah Jin and co-workers at the university of Colorado [147], who created a BEC of weakly bound molecules from ultra-cold Fermi gas of potassium-40 atoms. This new form of matter may help in a better understanding of superconductivity and could pave the way toward a superconductor that works at room temperature. Recently, research efforts in the field of BEC were directed toward storing ultra-cold bosonic and fermionic quantum gases in artificial periodic potentials of light, i.e., in optical lattices [148]. This has opened innovative manipulation and control possibilities, in many cases creating structures far beyond those currently achievable in typical condensed matter physics systems. Trapping ultra-cold quantum gases in optical lattices will open the door to a wide interdisciplinary field of physics ranging from non-linear dynamics to strongly correlated quantum phases and quantum information processing.

Around 2004, the group of atomic physics at the Physics Department of the Faculty of Science – Cairo University started theoretical work on BEC, and a brief review on the subject was given by El-Sherbini [149]. The group adopted in their studies the semiclassical approximation, i.e., the density of state (DOS) approach [150]. In this approach, the sum over the discrete spectrum for the thermodynamic quantities of the Bose–Einstein condensate was replaced by an integral weighted of a piecewise DOS. The latter was calculated via the technique of the high temperature expansion for the partition function [151]. This approximation has been widely used in variety of problems in statistical physics [152]. These studies showed that the resulting thermodynamic parameters depend crucially on the choice and construction of the DOS [153].

A generalization of the semiclassical approximation was suggested by Hassan and El-Badry [154], allowing for an essential extension of its region of applicability. The parameterized DOS has considered the effects of finite size, anisotropic of the harmonic potential, and the positive chemical potential; all of them simultaneously. The latter effect is similar to the effect of repulsive interaction provided by the mean field theory approach [155]. The outcome results provide a solid theoretical formulation for the existing experiments [155]. The generalized semiclassical approximation was also applied by Hassan and El-Badry [156] in the study of thermodynamic properties of quasi-equilibrium magnons in crystalline bulk materials and thin films and by Hassan [157] in the calculations of effective area and expansion energy of trapped Bose gas in a combined magnetic-optical potential. Hassan et al. [158] were able to calculate the critical temperature of a Bose–Einstein condensate in a 3D non-cubic optical lattice. Moreover, Hassan et al. [159] have studied recently the thermodynamic properties of condensed 39 K Bose gas in a harmonic trap. In particular the critical atoms number and its corresponding temperature are predicted via the graphical representation [160]. A step further in the advancement of research on BEC was carried out by the group, where they determined the thermodynamic properties of rotating Bose gas in a harmonic trap [161].

Nowadays, the work on fast rotating gases in Bose Einstein condensates which was initiated by the group of Dalibard at the Ecole Normale Superieure (ENS) laboratory in Paris [162] was recently under investigation by our group. Usually, the effective trapping potential V_{eff} for fast rotation regime is approximated by harmonic plus weak quartic potential [163]:

$$V_{eff} = \frac{m}{2}\left[\omega_z^2 z^2 + \omega_\perp^2 r_\perp^2\right] + \frac{1}{4}xr^4 - \frac{m}{2}\Omega^2 r_\perp^2 \tag{33}$$

where m is the mass of the atom and $r^2 = x^2 + y^2$ is the perpendicular radius to the rotation axis z. The parameter x classifies the strength of the quartic term and, moreover, plays an important role in exploring the fast rotation regime. The parameterκ must be greater than zero; otherwise, the effective potential felt by the atom V_{eff}would tend to $-\infty$ for $r \rightarrow \pm\infty$.

In a recent paper by the group, El-Sherbini et al. [164], a modified semiclassical approximation is provided to study the effects of finite size, the positive chemical potential, and anisotropy of the trap, on the thermodynamical properties of a rotating gas in a harmonic plus quartic trap. Fig. 15 presents the characteristic dependence of the condensate fraction N_0/N on the reduced temperature T/T_0 and the rotation rate α. It shows the monotonically decreasing nature of the condensate fraction due to the increase in the reduced temperature everywhere. This decrease is minor in the slow rotation range and rapid in the fast rotation range monotonically in agreement with the experimental observations and the numerical calculation [165]. Our results also provide a correction due to the finite size and positive chemical potential effects (interaction effect) for the results of Kling and Pelster [165]. Both of them show a significant quenching of the condensate fraction and a shift of the critical temperature toward the lower values.

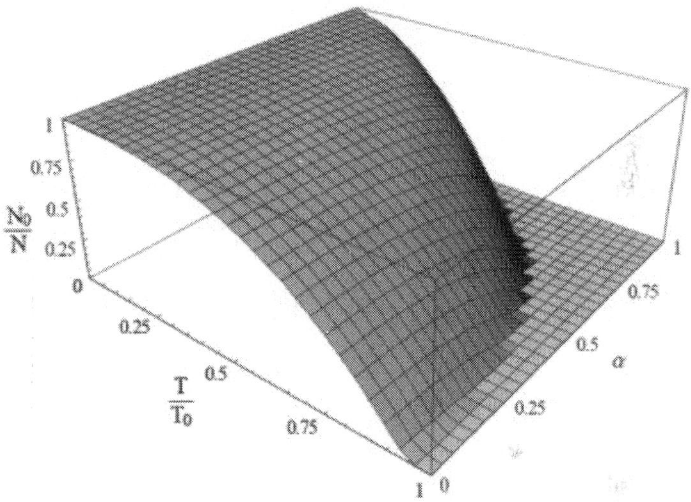

Figure 15. The condensate fraction N_0/N as a function of the reduced temperature T/T_0 and the rotation ratio α.

The study of BEC in a rotating optical lattice was done in the atomic physics group by Abdel-Gany et al. [166]. The effective trapping potential V_{eff} for rotating optical lattice trap is considered to be,

$$V_{eff} = \frac{m}{2}\left[\omega_z^2 z^2 + \omega_\perp^2 r_\perp^2\right] + V_0\left[sin^2\left(\frac{\pi x}{d}\right) + sin^2\left(\frac{\pi y}{d}\right)\right] - \frac{m}{2}\Omega^2 r_\perp^2 \tag{34}$$

where d is the lattice spacing. The dependence of the condensate fraction, critical temperature, and the heat capacity on the recoil frequency, optical potential depth, and rotating frequency is investigated. The results show that the normalized recoil frequency can control the characteristic shape of the condensate fraction Fig. 16. Furthermore, our results show that the rotating BEC in optical lattice is accompanied by a peak in the critical temperature T_c at defined value for the normalized recoil frequency $S_{\perp}$, Fig. 17. This remarkable behavior was observed experimentally by Burger et al. [167] and numerically detected for non-rotating boson gas in optical lattice by Blakie et al. [168].

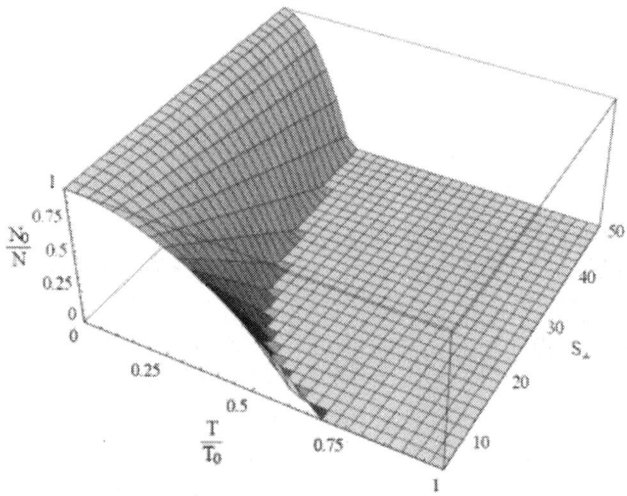

Figure 16. The condensate fraction N_0/N as a function of the normalized recoil frequency S and the reduced temperature T/T_0.

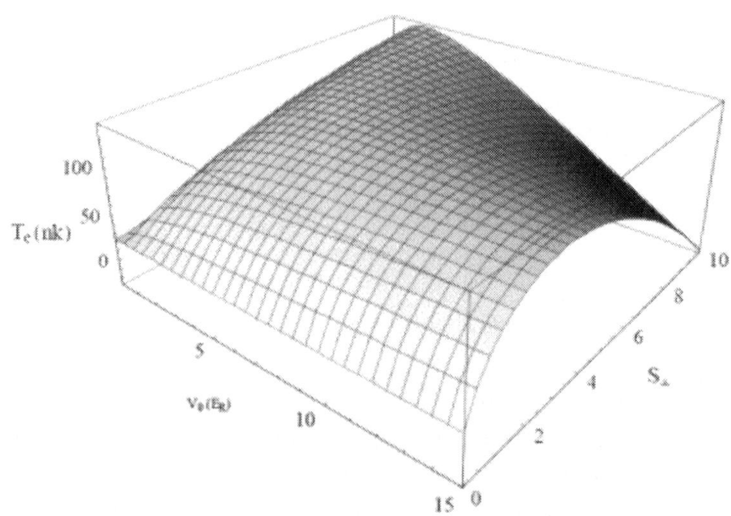

Figure 17. The normalized recoil frequency S as function of the critical temperature T_c and the optical depth V_0.

Bose–Einstein condensates in optical lattices represent model systems for solid state physics with yet unprecedented level of control. They can be used for exploring a wide range of fundamental problems in condensed matter physics such as Mott (metal-insulator) transitions, Anderson localization, type II superconductivity, and quantum Hall effect.

Exactly solvable models are very important in physics since they enable physicists to estimate the accuracy of the different approximate methods. In condensed atomic systems, the major problem involves solving many body interacting systems. The group of atomic physics at the Physics Department of the Faculty of Science – Cairo University has started to work on an exactly solvable model for a system composed of two species of identical Bosons in three-dimensional space interacting via harmonic potential. Aboul-Seoud et al. [169] have studied a system of two N-particles of identical Bosons with equal numbers, assuming that particles belonging to the same species repel each other and particles belong to different species attract each other. It was realized that the system is condensed in one channel when the coupling strengths are identical and in two channels when the coupling strengths are different. This study will enable us to understand the behavior of the miscibility of atomic species as a function of the variations in the coupling strengths between the Bose species.

Further research activities in the field of BEC are being pursued by members of our atomic physics group. The research focuses on the study of the thermodynamics of Bose systems at finite temperature and inter-particle interactions. A system of equations describes trapped and condensed Bose systems at finite temperature is being investigated. The description of weakly interacting Bosons at zero temperature saw a breakthrough in 1947 with the seminal work of Bogoliubov [170], who derived the mean field equations that govern the statics, as well as the dynamics, of the condensed state. Bogoliubov's work correctly describes the microscopic low energy spectrum of the condensed state of a uniform condensate. Further progress was achieved through the work of Gross and Pitaevskii [171] and [172], who derived the zero temperature equations of a non-uniform condensate. Their approach has become the basis for much of the theoretical work on degenerate Bose gases. This work saw a surge of activity after the experimental realization of Bose–Einstein Condensation in electromagnetic-optical traps containing metastable vapors of alkali atoms [144] and [145].

More work is needed, however, to address both dynamical and thermodynamic aspects of the condensed phase at finite temperatures. At finite temperature, the system contains bosonic quasi-particle excitations. As the temperature is raised, the density of such excitations increases and their interactions with the condensate particles, as well as among themselves, may have a pronounced effect on the physics of the trapped system. Another aspect comes about because of the experimental possibility to fine-tune the inter-particle interactions using Feschbach resonances. This means that the interactions between the atoms in the condensate are not weak anymore. The validity of many of the approaches that have been developed with both finite temperature and strong inter-particle interactions has been questioned by Yukalov [173]. In order to treat these effects consistently, he derived a system of coupled equations for the main quantities that govern the macroscopic properties of degenerate trapped gases. These quantities include the condensate density, the density of non-condensate particles, the local speed of sound, and the anomalous density of the trapped system. We are comparing the solutions we are obtaining with known experimental results [174]. This approach is promising since it appears that it can be easily generalized to include non-uniformities due to vortices or other special excitations, as well as dynamic phenomena at finite temperature. The results of our work are currently being prepared for submission for publication.

Concluding, I would like once more to stress that this review article was by no means an attempt to cover the whole of developments of atomic physics of the last half century. Rather, it was aimed at outlining the contributions of the atomic physics group of Cairo University that I had the honor to lead for the past four decades, as well as the research work I conducted during my scientific visits to the FOM-Institute for Atomic and Molecular Physics in Amsterdam.

With future publications, our group hopes to continue participating in the worldwide edifice of atomic physics.

Conflict of interest

The author has declared no conflict of interest.

Compliance with Ethics Requirements

This article does not contain any studies with human or animal subjects.

REFERENCES

1. Ramsauer C, Kollath R. Uber den wirkungsquerschnitt der edelgasmolek ule gegen uber elektronen unterhalb. Ann Phys 1929;3:536–64.
2. Ramsauer C, Kollath R. Die winkelverteilung bei der streuung langsamer elektronen an gasmoleku¨len. II. Fortsetzung. Ann Phys 1932;12:529–61.
3. Tate JT, Smith PT. Ionization potentials and probabilities for the formation of multiply charged ions in the alkali vapors and in krypton and xenon. Phys Rev 1934;46:773–6.
4. Bleakney W, Smith LG. The ionization probability of He++. Phys Rev 1936;49:402–12.
5. Hughes AL, Rojansky V. On the analysis of electronic velocities by electrostatic means. Phys Rev 1929;34:284–90
1. Phys Rev 1929;34:284..
6. Massey HSW, Smith RA. The passage of positive ions through gases. Proc Roy Soc 1933;A142:142–72.
7. Stueckelberg EC. Theorie der unelastischen Sto¨sse zwischen Atomen. Helv Phys Acta 1932;5:370–422.
8. Landau LD. On the theory of transfer of energy at collisions II. Z Phys Sov Un 1932;2:46–51.
9. Zener C. Non-adiabatic crossing of energy levels. Proc Roy Soc 1932;A137:696–702.
10. Massey HSW, Burhop EHS. Electronic and ionic impact phenomena. Oxford: The Clarendon Press; 1952.
11. Fox RE. Multiple ionization in argon and krypton by electron impact. J Chem Phys 1960;1960(33):200–5.
12. Fayard F. Energy dependence for single and double ionization by electron impact between 250 and 2200 eV. J Chem Phys 1968;48:478–87.
13. Van der Wiel MJ, EL-Sherbini ThM, Vriens L. Multiple ionization of He, Ne and Ar by 2–16 keV electrons. Physica 1969;42:411–20.

14. El-Sherbini ThM, Van der Wiel MJ, De Heer FJ. Multiple ionization of Kr and Xe by 2–14 keV electrons. Physica 1970;48:157–64.
15. Van der Wiel MJ. Small-angle scattering of 10 keV electrons in He, Ne, and Ar. Physica 1970;40:411–24.
16. Bethe HA, Rose ME, Smith LP. The multiple scattering of electrons. Proc Am Philos Soc 1938;78:573–85.
17. Krause MO, Carlson TA. Vacancy cascade in the irradiation of atoms with photons. Phys Rev 1967;158:18–24.
18. Cairns RB, Harrison H, Schoen RI. Multiple photoionization of xenon. Phys Rev 1969;183:52–9.
19. El-Sherbini ThM, Van der Wiel MJ. Oscillator strengths for multiple ionization in the outer and first inner shells of Kr and Xe. Physica 1972;62:119–38.
20. Samson JAR. The measurements of the photo-ionization cross sections of the atomic gases. Adv Atom Mol Phys 1966;2:177–260.
21. Amusia MYa, Cherepkov NA, Chernysheva LV. Cross sections for the photo-ionization of noble gas atoms with allowance of multi-electron correlations. Sov Phys – JETP 1971;33:90–6.
22. Wendin G. Collective resonance in the 4d10 shell in atomic Xe. Phys Lett A 1971;37:445–6.
23. Van der Wiel MJ, El-Sherbini ThM, Brion CE. K shell excitation of nitrogen and carbon monoxide by electron impact. Chem Phys Lett 1970;7:161–4.
24. El-Sherbini ThM, Van der Wiel MJ. Ionization of N2 and CO by 10 keV electrons as a function of the energy loss I. Valence electrons. Physica 1972;59:433–52.
25. Van der Wiel MJ, El-Sherbini ThM. Ionization of N2 and CO by 10 keV electrons as a function of the energy loss II. Innershell electrons. Physica 1972;59:453–62.
26. Bates DR. Atomic and molecular processes. London: Acad Press; 1962, p. 549.
27. Merzbacher E. Quantum mechanics. 2nd ed. John Wiley & Sons, Inc; 1970. p. 450–4.
28. Hasted JB. Inelastic collisions between ions and atoms. Proc Roy Soc Lond 1952;212A:235–48.
29. Hasted JB. Advances in electronics, vol. XIII. New York: Academic Press; 1960.
30. Morgan GH, Everhart E. Measurements of inelastic energy loss in large-angle Ar+ on Ar collisions at keV energies. Phys Rev 1962;128:667–76.
31. Kessel QC, Everhart E. Coincidence measurements of largeangle Ar+ on Ar collisions. Phys Rev 1966;146:16–27.

32. Niehaus A, Ruf M. An experimental study of autoionization processes occurring in 2 eV1 keV He2+–Hg collisions. I. Electron and ion energy spectra. J Phys B: Atom Mol Phys 1976;9:1401–18.
33. Winter H, Bloemen E, De Heer FJ. VUV radiation, slow ions and electrons produced in collisions of multiply charged Ne ions with He and Ar. J Phys B: Atom Mol Phys 1977;10: L599–604.
34. Winter H, El-Sherbini ThM, Bloemen E, De Heer FJ, Salop A. A comparison between radiative and non-radiative deexcitation after electron capture by multiply charged ions. Phys Lett 1978;68A:211–4.
35. Fano U, Lichten W. Interpretation of Ar+–Ar collisions at 50 KeV. Phys Rev Lett 1965;14:627–31.
36. Woerlee PH, El-Sherbini ThM, De Heer FJ, Saris FW. Energy spectra of electrons produced in collisions of multiply charged neon ions with noble-gas atoms. J Phys B: Atom Mol Phys 1979;12:L235–41.
37. Kishinevskii LM, Parilis ES. Auger ionization of atoms by mutiply-charged ions. Model of two coulomb centers. Sov Phys – JETP 1969;28:1020–34.
38. El-Sherbini ThM, Farrag A. Configuration interaction in the spectrum of Kr II. J Phys B: Atom Mol Phys 1976;9:2797–803.
39. El-Sherbini ThM, Salop A, Bloemen E, De Heer FJ. Target dependence of excitation resulting from electron capture in collisions of 200 keV Ar6+ ions with noble gases. J Phys B: Atom Mol Phys 1979;12:L579–82.
40. El-Sherbini ThM, Salop A, Bloemen E, De Heer FJ. Excitation and ionization resulting from electron capture in Ar6+ + H2 collisions at ion projectile energies of 200–1200 keV. J Phys B: Atom Mol Phys 1980;13:1433–49.
41. El-Sherbini ThM, ed. In: Proceedings of arab symposium on atomic and molecular processes in controlled thermonuclear fusion research, Cairo University 14–18 November, 1988.
42. El-Sherbini ThM, De Heer FJ. Projectile excitation in the collision of Arq+ (q = 1, 2 and 3) with He and Ne. J Phys B: Atom Mol Phys 1982;15:423–38.
43. Garstang RH. Some line strengths for ionized neon. Monthly Not Roy Astron Soc 1950;110:612–4.
44. Garstang RH. Transition probabilities of forbidden lines. J Res Nat Bur Stand, Sec A 1964;68:61–74.
45. Wiese WL, Smith MW, Miles BM. Atomic transition probabilities. 1969 Nat Bur Stand NSRDS-NBS4; 1969.
46. Luyken BFJ. Transition probabilities and radiative lifetimes for Ne II. Physica 1971;51:445–60.
47. Luyken BFJ. Transition probabilities and radiative lifetimes for Ar II. Physica 1972;60:432–66.

48. El-Sherbini ThM. Calculation of transition probabilities and radiative lifetimes for singly ionized krypton. J Phys B: Atom Mol Phys 1975;8:L183–4.
49. El-SherbiniThM.CalculationofXe IIline strengthsand radiative lifetimes in intermediate coupling. Z Phys 1975;A275:1–3.
50. El-Sherbini ThM. Line strengths and lifetimes for Kr II. Z Phys 1976;A276:325–7.
51. Aymar M, Crance M, Klapisch M. Results of parametric potential applied to rare gases. J Phys 1970;31:141–8.
52. Minnhagen L, Strihed H, Petarsson B. Revised and extended analysis of singly ionized krypton, Kr II. Ark Fys 1968;39:471–93.
53. El-Sherbini ThM, Zaki MA. Perturbations in the 5p4 6s and 5p4 5d configurations of Xe II. J Phys B: Atom Mol Phys 1978;11:2061–8.
54. El-Sherbini ThM. Transition probabilities and radiative lifetimes for singly ionized xenon. J Phys B: Atom Mol Phys 1976;9:1665–71.
55. El-Sherbini ThM. Configuration interaction in the spectrum of singly ionized xenon. In: Proceedings of the X international conference on the physics of electronic and atomic collisions (ICPEC), Paris 21–27 July 1977.
56. El-Sherbini ThM. Excitation mechanisms in the singly ionized krypton laser. Phys Lett 1982;88A:169–71.
57. Serrao JM. The absorption spectrum of beryllium. J Quant Spectrosc Radiat Transfer 1985;3:219–26.
58. El-Sherbini ThM, Mansour H, Farrag A. Hartree–Fock energies of the doubly excited states of the boron isoelectronic sequence. Ann Phys 1987;44:419–22.
59. Martinson I, Curtis LJ. Spectroscopic studies of the structure of highly ionized atoms. Cont Phys 1989;30:173–85.
60. Norreys P, Zhang J, Cairns G, Djaoui A, Dwivedi L, Key M, et al. Measurement of the photo-pump strength of the 3d–5f transitions in the automatically line matched Ni-like Sm photopumped X-ray laser. J Phys B: Atom Mol Opt Phys 1993;26:3693–9.
61. Nilson J. Lasing on the 3d–3p neon like X-ray laser transitions driven by a self-photo-pumping mechanism. Phys Rev A 1996;53:4539–47.
62. Mathews DL, Hagelstein PL, Rosen MD, Echart MJ, Ceglio NM, Hazi AU, et al. Demonstration of a soft X-ray amplifier. Phys Rev Lett 1985;54:110–3.
63. Hagelstein PL, Basv S, Muendel MH, Bravd JP, Tavber D, Kavshik S, et al. The MIT short-wavelength laser project: a status report. In: Proceedings of the international colloquium on X-ray laser, vol. 116, York, United Kingdom; 1990. p. 255–62.
64. Jaegle P, Jamelot G, Carillon A, Wehenkel C. Soft-X-ray amplification by lithiumlike ions in recombining hot plasmas. Jpn J Appl Phys 1987;17:563–74.

65. Silfvast WT, Wood OR. Photoionization lasers pumped by broadband soft-X-ray flux from laser-produced plasmas. J Opt Soc Am 1987;B4:609–18.
66. Jaegle P, Jamelot G, Carillon A, Klisnick A, Sureau A, Guennou H. Hot plasmas. J Opt Soc Am 1987;B4:563–74.
67. Krishnan M, Trebes J. Design considerations for optically pumped, quasi-cw, uv and xuv lasers in the Be isoelectronic sequence. Am Ins Phys 1984;46:514–27.
68. Feldman U, Seely JF, Bhatia AK. Density sensitive X-ray line ratios in the Be I, B I and Ne I isoelectronic sequences. J Appl Phys 1985;58:3953–7.
69. Al-Rabban MM. Electron excitation in the berylliumisoelectronic sequence and X-ray laser. MSc thesis, Faculty of Science, Ain Shams University; 1995.
70. Hibbert A. CIV3 – A general program to calculate configuration interaction wave functions and electric-dipole oscillator strengths. Comput Phys Commun 1975;9:141–72.
71. Friedrich H. Theoretical atomic physics. Springer Verlag; 1991.
72. Yahia ME, Azzouz IM, Allam SH, El-Sherbini ThM. Laser gain by electron collisional pumping of Ar VII–V XII. Opt Laser Technol 2008;40:1008–17.
73. Elton RC. X-ray lasers. Academic Press INC; 1990.
74. Vriens L, Smeets A. Cross-section and rate formulas for electron-impact ionization, excitation, deexcitat and total depopulation of excited atoms. Phys Rev 1980;A22:940–51.
75. Allam SH. CRMO-Collisional Radiative Model. Computer Code, Private Communication; 2005
76. Palumbo LJ, Elton RC. Short-wavelength laser calculations for electron pumping in carbonlike and heliumlike ions. J Opt Soc Am 1977;67:480–8.
77. El-Sherbini ThM, Wahby AS. Energy levels of the single excited states in He I isoelectronic sequence. Proc Ind Natl Sci Acad 1990;56A:39–46.
78. El-Sherbini ThM, Al-Rabban MM. Soft X-ray laser transitions in the Be-isoelectronic sequence. In: Proceedings of the 2nd international conference on lasers and applications, Cairo University 16–19 September 1996. p. 86–9.
79. El-Sherbini ThM, Farrag AA, Mansour HM, Rahman AA. Electrical dipole oscillator strengths and radiative lifetimes in the boron isoelectronic sequence. Ann Phys 1987;44:412–8.
80. El-Sherbini ThM, Mansour HM, Farrag AA, Rahman AA. Energy levels of the single excited states in the boron isoelectronic sequence. Ann Phys 1989;46:105–12.
81. El-Sherbini ThM, Al-Rabban MM. Laser transitions in the boron-like ions of NIII, OIV and FV. Infrared Phys 1991;31:595–7.
82. Allam SH, Farrag AA, Refaie AI, El-Sherbini ThM. Oscillator strengths of allowed transitions for CI-isoelectronic sequence. Arab J Nucl Sci Appl 1999;32:89–95.

83. El-Sherbini ThM, Farrag AA. Core-excited doublet and quartet states in the sodium isoelectronic sequence. J Quant Spectrosc Radiat Transfer 1991;46:473–5.
84. Younis WO, Allam SH, El-Sherbini ThM. Fine-structure calculations of energy levels, oscillator strengths, and transition probabilities for sodium-like ions (Co XVII–Kr XXVI). Atom Nucl Data Tables 2006;92:187–205.
85. Younis WO, Allam SH, El-Sherbini ThM. Rate coefficients for electron impact excitation, de-excitation and laser gain calculations of the excited ions Co XVII up to Br XXV. Can J Phys 2010;88:257–69.
86. El-Sherbini ThM, Rahman AA. Auto-ionizing states in Mg I. Ann Phys 1982;39:333–7.
87. El-Sherbini ThM, Mansour HM, Farrag AA, Rahman AA. Electric dipole oscillator strength and radiative lifetimes in the magnesium isoelectronic sequence. Ann Phys 1988;45:498–506.
88. El-Sherbini ThM, Mansour HM, Farrag AA, Rahman AA. Energy levels of the single excited states in the magnesium isoelectronic sequence. Ann Phys 1989;46:144–8.
89. Younis WO, Allam SH. Relativistic energy levels and transition probabilities for Al-like ions (Z = 33–35). Int Rev Phys 2011;5:207–46.
90. El-Maaref AA, Uosif MA, Allam SH, El-Sherbini MTh. Energy levels, oscillator strengths and transition probabilities for Si-like P II, S III, Cl IV, Ar V and K VI. Atom Data Nucl Data Tables 2012;98:589–615.
91. El-Maaref AA, Allam SH, Uosif MA, El-Sherbini ThM. The 4d–4p transition and soft X-ray laser wavelengths in Si-like ions. Can J Phys 2013;91:1–13.
92. El-Maaref AA, Allam SH, El-Sherbini ThM. Energy levels, oscillator strengths and radiative rates of Si-like Zn XVII, Ga XVIII, Ge XIX and As XX. Atom Data Nucl Data Tables 2013
2. in press..
93. El-Maaref AA, Ahmed M, Allam SH. Fine-structure calculations of energy levels, oscillator strengths and transition probabilities for sulfur-like Fe XI. Atom Data Nucl Data Tables 2013
94. Lamia MA. Laser transitions in neutral potassium and potassium like ions, MSc thesis, Faculty of Science, Benha University; 2009.
95. Refaie AI, El-Sharkawy H, Allam SH, El-Sherbini ThM. Theoretical electron impact excitation, ionization and recombination rate coefficients and level population densities for scandium-like ions. Int J Pure Appl Phys 2007;3:75–82.
96. Abdelaziz WS, El-Sherbini ThM. Reduced population and gain coefficient calculations for soft X-ray laser emission from Eu35+. Opt Laser Technol 2010;42:699–702.

97. Tognoni E, Palleschi V, Corsi M, Cristoforetti G. Quantitative micro-analysis by laser-induced breakdown spectroscopy: a review of the experimental approaches. Spectrochim Acta 2002;B57:1115–30.
98. Barrette L, Turmel S. On-line iron-ore slurry monitoring for real-time process control of pellet making processes using laserinduced breakdown spectroscopy: graphitic vs. total carbon detection. Spectrochim Acta 2001;B56:715–23.
99. Hassan M, Sighicelli M, Lai A, Colao F, Ahmed AH, Fantoni R, et al. Studying the enhanced phytoremediation of lead contaminated soils via laser induced breakdown spectroscopy. Spectrochem Acta 2008;B63:1225–9.
100. Abdel-Salam Z, Al Sharnoubi J, Harith MA. Qualitative evaluation of maternal milk and commercial infant formulas via LIBS. Talanta 2013;115:422–6.
101. St-Onge L, Kwong E, Sabsabi M, Vadas E. Rapid analysis of liquid formulations containing sodium chloride using laser-induced breakdown spectroscopy. J Pharm Biomed Anal 2004;36:277–84.
102. Narayanan V, Thareja R. Emission spectroscopy of laserablated Si plasma related to nanoparticle formation. Appl Surf Sci 2004;222:382–93.
103. Garc'ya-Ayuso L, Amador-Hernandez J, Fernandez-Romero J, De Castro M. Characterization of jewellery products by laser-induced breakdown spectroscopy. Anal Chim Acta 2002;457:247–56.
104. Capitelli F, Colao F, Provenzano M, Fantoni R, Brunetti G, Senesi N. Determination of heavy metals in soils by laser induced breakdown spectroscopy. Geoderma 2002;106:45–62.
105. Galmed AH, Kassem AK, Von Bergmann H, Harith MA. A study of using femtosecond LIBS in analyzing metallic thin film – semiconductor interface. Appl Phys 2011;B102:197–204.
106. Khedr A, Papadakis V, Pouli P, Anglos D, Harith MA. The potential use of plume imaging for real-time monitoring of laser ablation cleaning of stonework. Appl Phys 2011;B105:485–92.
107. Colao F, Fantonia R, Lazica V, Paolinia A, Fabbria F, Orib G, et al. LIBS feasibility for in situ planetary exploration: an analysis on Martian rock analogues. Planet Space Sci 2004;52:117–23.
108. Sanchez-Ake C, Bolanos M, Ramirez C. Emission enhancement using two orthogonal targets in double pulse laser-induced breakdown spectroscopy. Spectrochim Acta Part B: Atom Spectrosc 2009;64:857–62.
109. Cristoforetti G, Legnaioli S, Pardini L, Palleschi V, Salvetti A, Tognoni E. Spectroscopic and shadow-graphic analysis of laser induced plasmas in the orthogonal double pulse pre-ablation configuration. Spectrochim Acta Part B: Atom Spectrosc 2006;61:340–50.

110. Piepmeier EH, Malmstadt HV. Q-switched laser energy absorption in the plume of an aluminum alloy. Anal Chem 1969;41:700–7.
111. Sattmann R, Sturm V, Noll R. Laser-induced breakdown spectroscopy of steel samples using multiple Q-switch Nd:YAG laser pulses. J Phys D: Appl Phys 1995;28:2181–7.
112. Petukh ML, Rozantsev VA, Shirokanov AD, Yankovskii AA. The spectral intensity of the plasma of single and double laser pulses. J Appl Spectrosc 2000;67:1097–101.
113. St-Onge L, Detalle V, Sabsabi M. Enhanced laser-induced breakdown spectroscopy using combination of fourthharmonic and fundamental Nd:YAG pulses. Spectrochim Acta B: Atom Spectrosc 2002;57:121–35.
114. Noll R. Laser induced breakdown spectroscopy. BerlinHeidelberg: Springer Verlag; 2012.
115. Tarasenko NV, Burakov VS, Butsen AV. Laser ablation plasmas in liquids for fabrication of nanosize particles. In: VI serbian–belarusian symp on phys and diagn of lab & astrophys plasma, vol. 82, Belgrade, Serbia, 22–25 August 2006, Pub Astron Obs Belgrade, 2007. p. 201–11.
116. Griem HR. Plasma spectroscopy. McGraw-Hill Inc.; 1964.
117. Griem HR. Spectral line broadening by plasmas. New York: Academic Press; 1974.
118. Hegazy H, Sharkawy H, El-Sherbini ThM. Use of spectral lines of pure Ti target for the spectroscopic diagnostics of the laser induced plasma in vacuum. Proc AIP Conf 2007;888:152–9.
119. El-Sherbini AM, Abdel Hamid A, El-Sherbini ThM. Collisionalradiative model for neutral argon plasma. In: Proceedings of the XXVth international conference on solid state physics and materials science, Luxor 6–10 March, 2005. p. 14–18.
120. El-Sergany F, Atta Khedr M, Abuzeid H, Montasir M, ElSherbini ThM. Elemental analysis of some minerals using the laser induced plasma spectroscopy. In: Proceedings of the first Euro-mediter symposium, Cairo 2–6 November, 2000. p. 14–7.
121. El-Sherbini AM, Hegazy H, El-Sherbini ThM, Cristoforretti G, Legnaioli S, Palleschi V, et al. Evaluation of self-absorption coefficients of aluminum emission lines in laser-induced breakdown spectroscopy measurements. Spectrochim Acta Part B: Atom Spectrosc 2005;60:1573–9.
122. El-Sherbini AM, Hegazy H, El-Sherbini ThM. Measurement of electron density utilizing the Ha-line from laser produced plasma in air. Spectrochim Acta Part B: Atom Spectrosc 2006;61:532–9.

123. El-Sherbini AM, El-Sherbini ThM, Hegazy, Cristoforetti G, Leganaioli S, Pardini L, et al. Measurement of the Stark broadening coefficient of atomic emission lines by laserinduced breakdown spectroscopy technique. Spectrosc Lett 2007;40:643–58.
124. El-Sherbini AM, Refaie AI, Aboulfotouh AM, El-Kamhawy AA, Abdel Sabour K, Imam H, et al. An assessment of LIBSdiagnostics of plasma using the hydrogen Ha-line at different laser energies. In: Proceedings of the third arab international conference on physics and materials science, Alexandria 21–23 October, 2009. p. 176–87.
125. El-Sherbini AM, Aboulfotouh AM, Allam SH, El-Sherbini ThM. Diode laser absorption measurements at the Hatransition in laser induced plasmas on different targets. Spectrochim Acta Part B: Atom Spectrosc 2010;65:1041–6.
126. El-Sherbini AM, Al Amer AS, Hassan AT, El-Sherbini ThM. Measurements of plasma electron temperature utilizing magnesium lines appeared in laser produced aluminum plasma in air. Opt Photon J 2012;2:278–85.
127. El-Sherbini AM, Al Amer AS, Hassan AT, El-Sherbini ThM. Spectrometric measurement of plasma parameters utilizing the target ambient gas O I and N I atomic lines in LIBS experiment. Opt Photon J 2012;2:286–93.
128. Hegazy H, Allam SH, Chaurasia S, Dhareshwar L, El-Sherbini ThM, Kunze HJ, et al. Joint experiments on X-ray/particle emission from plasmas produced by laser irradiating nano structured targets. Am Inst Phys Plasma Fusion Sci 2008;996: 243–50.
129. El-Sherbini AM, Aboulfotouh AM, Rashid FF, Allam SH, ElDakrouri A, El-Sherbini ThM. Observed enhancement in LIBS signals from nano vs. bulk targets: comparative study of plasma parameters. World J Nano Sci Eng 2012;2:181–8.
130. Hegazy H, Abdel-Rahim FM, Allam SH. Evolution of Al plasma generated by Nd-Yag laser radiation at the fundamental wavelength. Appl Phys B 2012;108:665–73.
131. Hegazy H, Abdel-Ghany HA, Allam SH, El-Sherbini ThM. Spectral evolution of nano-second laser interaction with Ti target in air. Appl Phys B 2013;110:509–18.
132. Hermann J, Thomann AL, Boulmer-Leborgne C, Dubereuil B, De Giorgi ML, Perrone A, et al. Plasma diagnostics in pulsed laser TiN layer deposition. J Appl Phys 1995;77:2928–36.
133. Einstein A. On the quantum theory of radiation. Phys Z 1917;18:121–8.
134. Phillips W, Metcalf H. Laser deceleration of an atomic beam. Phys Rev Lett 1982;48:596–9.
135. Prentiss M, Cable A. Slowing and cooling an atomic beam using an intense optical standing wave. Phys Rev Lett 1989;62:1354–7.

136. Metcalf H, Van der Straten P. Laser cooling and trapping of neutral atoms. Phys Rep 1994;244:203–86.
137. Prodan J. Stopping atoms with laser light. Phys Rev Lett 1985;54:992–5.
138. Hansch TW, Schawlow AL. Cooling of gases by laser radiation. Opt Commun 1975;13:68–9.
139. Chu S, Hollberg L, Bjorkholm JE, Cable A, Ashkin A. Threedimensional viscous confinement and cooling of atoms by resonance radiation pressure. Phys Rev Lett 1985;55:48–51.
140. Pethick CJ, Smith R. Bose–Einstein condensation in dilute gases. Cambridge University Press; 2002.
141. Hess HF. Evaporative cooling of magnetically trapped and compressed spin-polarized hydrogen. Phys Rev B 1986;34:3476–9.
142. Ketterle W. When atoms behave as waves: Bose–Einstein condensation and the atom laser. Rev Mod Phys 2002;74: 1131–51.
143. Molenear PA, Van der Straten P, Heideman HG. Diagnostic technique for Zeeman-compensated atomic beam slowing: technique and results. Phys Rev A 1997;55:605–14.
144. Anderson MH, Ensher JR, Matthews MR, Wieman CE, Cornell EA. Observation of Bose–Einstein condensation in a dilute atomic vapor. Science 1995;269:198–201.
145. Davis KB, Mewes MO, Andrews MR, Van Druten NJ, Durfee DS, Kurn DM, et al. Bose–Einstein condensation in a gas of sodium atoms. Phys Rev Lett 1995;75:3969–73.
146. Bradley CC, SAckett CA, Hulet RG. Bose–Einstein condensation of lithium: observation of limited condensate number. Phys Rev Lett 1997;78:985–9.
147. Greiner M, Regal C, Jin D. Emergence of a molecular Bose– Einstein condensate from a Fermi gas. Nature 2003;426: 537–40.
148. Bloch I. Ultracold quantum gases in optical lattices. Nat Phys 2005;1:23–30.
149. El-Sherbini ThM. Bose–Einstein condensation. Proc Am Inst Phys Conf 2004;748:44–54.
150. Bagnato V, Pritchard DE, Kleppner D. Bose–Einstein condensation in an external potential. Phys Rev A 1987;35: 4354–8.
151. Ketterle W, Van Druten NJ. Bose–Einstein condensation of a finite number of particles trapped in one or three dimensions. Phys Rev A 1996;54:656–60.
152. Dalfovo F, Giorgini S, Pitaevskii LP, Stringari S. Theory of Bose–Einstein condensation in trapped gases. Rev Mod Phys 1999;71:463–512.
153. Kirsten K, Toms DJ. Bose–Einstein condensation in arbitrarily shaped cavities. Phys Rev E 1999;59:158–67.
154. Hassan AS, El-Badry AM. Critical points of a three-dimensional harmonically trapped Bose gas. Physica B 2009;404:1947–50.

155. Williams RA, Al-Assam S, Foot CJ. Observation of vortex nucleation in a rotating two-dimensional lattice of Bose– Einstein condensates. Phys Rev Lett 2010;104:050404–7; Smith RP, Campbell RL, Tammuz N, Hadzibabic Z. Can a Bose gas be saturated? Phys Rev Lett 2011;106:230401–4.
156. Hassan AS, El-Badry AM. Effective width and expansion energy of the interacting condensed 87Rb Bose gas with finite size effects. Turk J Phys 2009;33:21–30.
157. Hassan AS. Effective area and expansion energy of trapped Bose gas in a combined magnetic–optical potential. Phys Lett A 2010;374:2106–12.
158. Hassan AS, El-Badry AM, Soliman SS. Critical temperature of a Bose–Einstein condensate in a 3D non-cubic optical lattice. Physica B 2010;405:4768–71.
159. Hassan AS, El-Badry AM, Soliman SS. Thermodynamic properties of a condensed Bose gas in a harmonic trap. Physica B 2013;410:63–7.
160. Hassan AS, El-Badry AM, Soliman SS. Thermodynamic properties of a rotating Bose gas in harmonic trap. Eur Phys J D 2011;64:465–71.
161. Hassan AS, El-Badry AM, Mohammedein AM, Ebeid MR. Effective widths of boson gas confined in a harmonic rotating trap. Phys Lett A 2012;376:1781–5.
162. Bretin V, Stock S, Seurin Y, Dalibard J. Fast rotation of a Bose–Einstein condensate. Phys Rev Lett 2004;92:050403–6.
163. Blanc X, Rougerie N. Lowest Landau level vortex structure of a Bose–Einstein condensate rotating in a harmonic plus quartic trap. Phys Rev A 2008;77:053615–22.
164. El-Sherbini ThM, Hassan D, Galal AA, Hassan AS. Thermodynamic properties of a rotating Bose-Einstein condensa tion in a harmonic plus quartic trap. Eur Phys J 2013;67:185–91.
165. Kling S, Pelster A. Thermodynamical properties of a rotating ideal Bose gas. Phys Rev A 2007;76:023609–14.
166. Abdel-Gany HA, Ellithi AY, Galal AA, Hassan AS. Thermodynamic properties of a rotating Bose–Einstein condensation in a 2D optical lattice. Tuk J Phys 2014;38:39–49.
167. Burger S, Cataliotti FS, Fort C, Madaloni P, Minardi F, Inguscio M. Quasi-2D Bose–Einstein condensation in an optical lattice. Eur Phys Lett 2002;57:1–6.
168. Blakie PB, Bezett A, Buonsante PF. Degenerate Fermi gas in a combined harmonic-lattice potential. Phys Rev A 2007;75:053609–18.
169. Aboul-Seod A, Hussein A, Galal AA, El-Sherbini ThM. An exactly solvable model of two species of identical bosons interacting via a harmonic oscillator potential. Phys Rev A 2013

170. Bogoliubov NN. On the theory of superfluidity. J Phys USSR 1947;11:23–32.
171. Gross EP. Structure of a quantized vortex in boson systems. Nuovo Cimento 1961;20:454–7.
172. Pitaevskii LP. Vortex lines in an imperfect Bose gas. Sov Phys JETP 1961;13:451–4.
173. Yukalov VI. Bose–Einstein condensation and gauge symmetry breaking. Laser Phys Lett 2007;4:632–47.
174. Gerbier F, Thywissen JH, Richard S, Hugbart M, Bouyer P, Aspect A. Critical temperature of a trapped, weakly interacting Bose gas. Phys Rev Lett 2004;92:030405–8.

CITATION

Tharwat M. El-Sherbini, Advances in atomic physics: Four decades of contribution of the Cairo University – Atomic Physics Group, Journal of Advanced Research, Volume 6, Issue 5, September 2015, Pages 643-661, ISSN 2090-1232, http://dx.doi.org/10.1016/j.jare.2013.08.004.

CHAPTER 5

A Modern Perspective on the History of Semiconductor Nitride Blue Light Sources

Herbert Paul Maruska, Walden Clark Rhines

Mentor Graphics, 8005 SW Broeckman Road, Wilsonville, OR 97070, United States

ABSTRACT

In this paper we shall discuss the development of blue light-emitting (LED) and laser diodes (LD), starting early in the 20th century. Various materials systems were investigated, but in the end, the nitrides of aluminum, gallium and indium proved to be the most effective. Single crystal thin films of GaN first emerged in 1968. Blue light-emitting diodes were first reported in 1971. Devices grown in the 1970s were prepared by the halide transport method, and were never efficient enough for commercial products due to contamination. Devices created by metal–organic vapor-phase epitaxy gave far superior performance. Actual true blue LEDs based on direct band-to-band transitions, free of recombination through deep levels, were finally developed in 1994, leading to a breakthrough in LED performance, as well as nitride based laser diodes in 1996. In 2014, the scientists who achieved these critical results were awarded the Nobel Prize in Physics.

HIGHLIGHTS

- This review discusses the history of blue light-emitting diodes since 1923.
- Particular attention is placed on the slow but steady improvement in device efficiency since the 1970s.
- Work in the Japanese laboratories where the problems with blue device performance were finally solved in 1996 is emphasized.
- The discussion concludes with the award of the Nobel Prize in Physics to Nakamura, Akasaki, and Amano in December, 2014.

EARLY HISTORY OF SEMICONDUCTOR LIGHT EMITTERS

Starting early in the 20th century, there were several reports of light emission from materials due to applied electric fields, a phenomenon which was termed "electroluminescence." The materials properties were poorly controlled, and the emission processes were not well understood. For example, the first report in 1923 of blue electroluminescence was based on light emission from particles of SiC which had been manufactured as sandpaper grit, and which contained accidental p–n junctions. By the late 1960s, SiC films had been prepared by more careful processes, and p–n junction devices were fabricated, leading to the first blue light-emitting diodes (LEDs) [1]. Electrical to optical conversion efficiencies were never more than about 0.005% [2]. In the ensuing decades, blue SiC LEDs were never substantially improved, because SiC has an indirect band gap. Although many blue SiC LEDs were actually sold commercially in the early 1990s, they are no longer a viable product. In the end, the best SiC LEDs, emitting blue light at 470 nm, had an efficiency of only 0.03% [3].

Starting in the late 1950s, various research laboratories such as IBM, RCA, and GE looked into the possibility of making infrared Light-Emitting Diodes, followed by visible light LEDs based on the III–V compound semiconductors. For example, at RCA Laboratories, Jacques Pankove reported infrared electroluminescence from GaAs in 1962 [4]. The beginning of visible-spectrum LEDs made of ternary III–V alloy compounds dates back to 1962 when Holonyak and Bevacqua [5] (1962) reported on the emission of visible light from GaAsP junctions in the first volume of *Applied Physics Letters*. After significant efforts, in 1964 green GaP LEDs were formed with efficiencies as high as 0.6% by doping the GaP with N isoelectronic impurities [6]. The N was added in the form of GaN particles to the growth melts used to form the p–n junctions. Improvements continued in the next several years for GaP green LEDs, and by 1968 these became commercial products [7]. While the external quantum efficiency of green LEDs is less than that of red LEDs, the sensitivity of the human eye to green light is more than 10 times higher than to red light, so the apparent brightness of the LEDs is comparable. But extending the results to blue LEDs presented a new challenge.

Both SiC and GaP have indirect band gaps. That means that an electron in the conduction band and a hole in the valence band have different values of momentum. To conserve momentum, a phonon must also be involved in the recombination process. This 3-body event suffers from low probability, and hence indirect band gap materials tend to have low emission efficiencies. But the presence of impurities in the semiconductor can relieve the conservation of momentum requirement by relying on Heisenberg's Uncertainty Principle. Since the physical position of the impurity is precisely specified, the momentum of a carrier in its vicinity becomes unspecified. This argument has led to commercial products with both GaP and SiC.

NITRIDE SEMICONDUCTORS

Gallium nitride films finally emerged out of the laboratories of the Radio Corporation of America (RCA) starting in 1968. RCA was an important center for compound semiconductor research. Starting in the mid-1950s, managers of the firm's research laboratories in Princeton, New Jersey made significant investments in compound semiconductors. They recruited young PhDs to the compound semiconductor program and reassigned experienced researchers who had previously worked on silicon and germanium. By the late 1960s, RCA had one of the largest research groups on compound semiconductors in the United States with thirty PhD researchers and more than fifty technicians and associate researchers.

Then, in 1965 and 1966, two young PhD chemists at RCA, James J. Tietjen and James A. Amick, developed a new epitaxial method for fabricating compound semiconductor crystals: halide vapor phase epitaxy or HVPE. Unlike liquid phase epitaxy, HVPE relied on the chemical reaction of different gases, including hydrides (hydrogen-based compounds) and halides (hydrogen chloride). HVPE had the advantage of offering more control of crystal growth than liquid phase epitaxy. The new crystal growing techniques enabled the firm's researchers to fabricate a large number of compound semiconductor materials. They developed competencies in indium phosphide, gallium arsenide, and various gallium arsenide alloys, including gallium arsenide phosphide. [8] Soon researchers at RCA and several other laboratories succeeded in fabricating red and then green LEDs.

In the spring of 1968, Tietjen became interested in using his HVPE method to grow another III–V compound semiconductor material: single crystal films of gallium nitride (no other laboratory had yet produced such crystals). Tietjen thought that the growth of gallium nitride crystals would make it possible to fabricate blue LEDs. He surmised that gallium nitride, because of the position of gallium and nitrogen in the periodic table of elements, would have one of the widest band gaps among compound semiconductors. This wide band gap would make gallium nitride a valuable semiconductor for blue light emission. Tietjen further reasoned that gallium nitride-based blue LEDs would have great commercial potential. The compound semiconductor group at RCA already knew how to make infrared, red, and green LEDs, so advancing to blue devices seemed to provide no special challenge. Adding blue would make flat panel televisions possible and end the era of the venerable cathode ray tube.

In the late 1960s Herbert Maruska was a young researcher in Tietjen's group and Tietjen assigned him to develop a process based on HVPE for growing thin films of gallium nitride. Tietjen and Maruska settled on sapphire as the crystal substrate to grow gallium nitride; this was a prescient choice as the great majority of gallium nitride crystals have been grown on sapphire up to this day. Maruska then struggled for many months to find the right temperature for growing gallium nitride films. At first, he used low temperatures, in the order of 600 °C, to grow the crystals, because he misinterpreted the earlier results of Lorenz and Binkowski [9]. He obtained only white powder. It was only by raising the temperature to 900 °C, the temperature commonly used for growing gallium arsenide, that he succeeded in making single crystal films of gallium nitride. This important breakthrough was accomplished on November 22, 1968.

The GaN films always came with high n-type conductivity. In the succeeding months, Maruska studied a variety of possible p-type dopants. But he did not succeed in making a standard pn junction LED. Traditionally, pn junction LEDs were made of a sandwich of n-type and p-type layers, as exhibited in red GaAsP and green GaP devices. In fact, it would prove to be exceedingly difficult to make p-type gallium nitride over the next 20 years. In 1969, he tried to dope gallium nitride with zinc, the standard element which provides p-type doping in GaAsP and GaP. He also used magnesium, cadmium, mercury, and germanium. None of these efforts succeeded, however, because he had run into the major doping problem with gallium nitride that would plague researchers aiming at making blue LEDs for the next twenty years: poor crystal quality and impurity contamination. Tietjen and Maruska published an article on the growth of single crystals of gallium nitride in *Applied Physics Letters* in November 1969 [10].

In early 1970, Jacques Pankove, an advanced researcher in the field of compound semiconductors, returned to RCA Laboratories from a sabbatical at UC Berkeley where he had written a textbook on optical processes in semiconductors. He soon joined the nitride research group to study gallium nitride. So did Edward Miller, a research chemist. Meanwhile, in March 1970, Maruska was awarded a David Sarnoff Doctoral Study Award. The Lab Director, Dr. Fred Rosi, assigned him to attend Stanford University and develop a bright blue GaN LED as his thesis project. In the fall of 1970, Maruska left to join the materials science and engineering department at Stanford University to commence his doctoral studies. At Stanford, he continued working on gallium nitride device structures with RCA funding and in close collaboration with Pankove. He remained an RCA employee.

The prospect of commercial opportunities encouraged several electronics corporations to build research groups focusing on gallium nitride in the early 1970s. Among these firms were large U.S. corporations with significant research operations such as North American-Rockwell, Phillips Laboratories, and the Bell Telephone Laboratories. Several Japanese corporations such as Hitachi and Toshiba also followed the trend. Most researchers working on gallium nitride at these firms followed the RCA example and grew their films of gallium nitride with the HVPE method.

GALLIUM NITRIDE LIGHT-EMITTING DIODES

When Maruska headed west to Stanford in September 1970, Edward Miller, who was in Dave Richman's group, took over the growth of the gallium nitride films. He thus became Jacques Pankove's collaborator, growing the films and passing them along to Pankove's lab to fabricate devices and evaluate them. Miller originally had focused his efforts on trying to prepare conducting p-type GaN. In the spring of 1971, Pankove took one of Miller's Zn-doped samples and placed two probes on the surface. He then applied a dc voltage source between the probes. When the voltage exceeded about 60 V, green light at 515 nm was emitted in the vicinity of one of the probes (Fig. 1). A microscopic investigation indicated that the luminescence consisted of a number of tiny spots located under the metal probe. This proved to be the world's first report of electroluminescence from a GaN film. In May 1971, this result was submitted for publication [11].

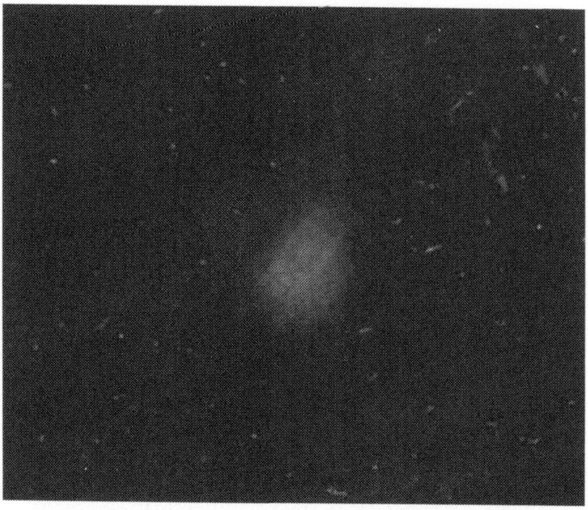

Figure 1. Ed Miller's very first GaN green LED [11].

Soon Miller fabricated a blue LED (475 nm) as well (Fig. 2) [12]. Pankove applied for a patent for GaN LEDs in 1971, and it was issued in 1972 [13]. Pankove and Miller went on to produce a blue GaN numeric display based on Zn-doped GaN in November 1972 (Fig. 3). Pankove and Miller published numerous papers about their Zn-doped GaN LEDs in the next few years [14], [15] and [16].

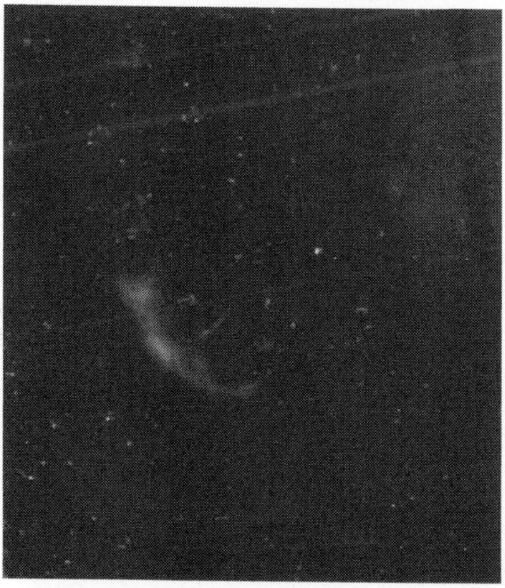

Figure 2. Ed Miller's first Zn-doped GaN blue LED [12].

Figure 3. GaN numeric display [14].

Early in the summer of 1971, Maruska requested and was sent all of the proprietary mechanical drawings for Tietjen's HVPE reactor from RCA, where he was still an employee. The drawings were taken to the machine shop in the McCullough Building at Stanford, and soon the technicians were machining the necessary parts for the HVPE reactor. As the summer of 1971 went by and the parts were completed, the reactor was slowly assembled. In addition to the parts which were fabricated in the machine shop, many other pieces of equipment needed to be ordered from outside vendors. These included the controllers which were used for setting the temperatures of the furnaces, and reactor gases including a cylinder of hydrogen chloride, a cylinder of ammonia, and tanks of hydrogen. The hydrogen needed to be purified with a palladium diffuser. The reactor is shown below (Fig. 4).

Figure 4. RCA-style HVPE reactor for GaN growth at Stanford University.

After Maruska had passed his doctoral qualifying examination on April 6, 1972, he was excited to get down to work and start to operate the new HVPE reactor. Naturally the reactor first needed to go on a shake-down cruise. This meant growing simple unintentionally doped material. By the middle of May, nice films of undoped GaN were being prepared. Next the use of zinc was commenced to form a thin doped layer above the thick undoped film, and it was not long before a green point-contact LED had been demonstrated. With a +70 V bias applied, green light peaking at 515 nm was observed in the vicinity of the positive probe. This was in agreement with the results of Miller and Pankove.

Mg-DOPED BLUE GaN LED

Aware that a PhD thesis has to be based on original research, Maruska realized that he needed to adopt a different approach compared to his colleagues back East. He spoke at length with another graduate student, Walden C. Rhines, who sat at an adjacent desk in the McCullough Building at Stanford. They decided to substitute magnesium for zinc to create a novel acceptor. Soon they were preparing GaN films doped with magnesium which were pale yellow in color and electrically insulating. In early June of 1972, several device structures were prepared based on a thick undoped (n-type) GaN film topped with a thin Mg-doped insulating film. Point contact probes were used for providing electrical contact. On June 8, 1972, after applying 150 V to a set of point contacts, the first example of violet light emission from Mg-doped GaN was observed. The emission peaked at 425 nm, which is indeed in the violet region of the visible spectrum. The human eye has very low sensitivity in this wavelength range, so such devices did not appear to be very bright, and in fact, looked blue. But they could be easily seen even with the room lights on.

It was now time to make a functioning metal–insulator–semiconductor device. A schematic diagram is shown below. Typically a chip about 2 mm by 2 mm was cut from the sample. The top surface of the Mg-doped film was then coated with a large metal contact. A small metal contact was also formed on the side of the chip to make contact to the undoped GaN layer. An indium amalgam typically used to make contact to conventional GaAs devices was used for this purpose. The little diode was then placed upside down in a standard TO-5 header, where the large metal region over the Mg-doped film made direct contact with the metal of the header. A separate wire passed through the header from the back and contacted the undoped region from the side. The device was heated to 400 °C to drive off the mercury and provide a solid metal contact. Thus all of these early samples were subjected to a thermal annealing procedure. With a positive bias applied to the header base, light was emitted out of the top, through the sapphire substrate. Rhines and Maruska published the first paper on Mg-doped blue LEDs in June 1972 [17]. See Fig. 5 below.

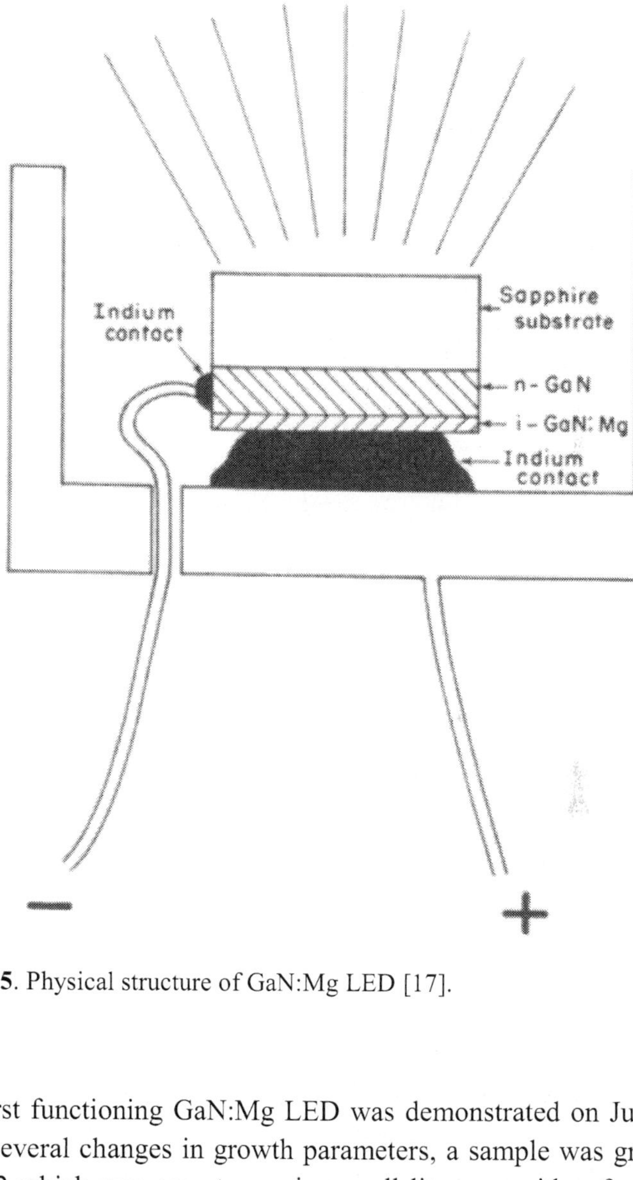

Figure 5. Physical structure of GaN:Mg LED [17].

The first functioning GaN:Mg LED was demonstrated on June 28, 1972. After several changes in growth parameters, a sample was grown on July 7, 1972 which was easy to see in a well lit room with a forward bias of 10 V [18]. See Fig. 6 below. Magnesium doping of gallium nitride would become the standard for future bright blue light emitters (see Fig. 7 and Fig. 8).

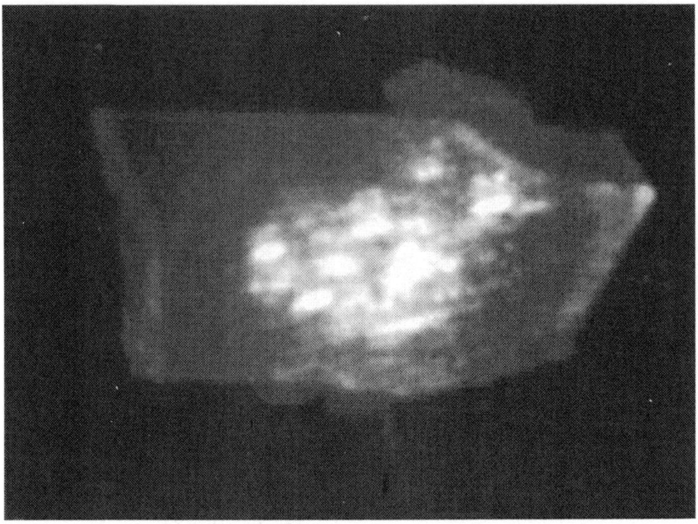

Figure 6. Blue GaN:Mg LED prepared at Stanford on July 7, 1972 [18].

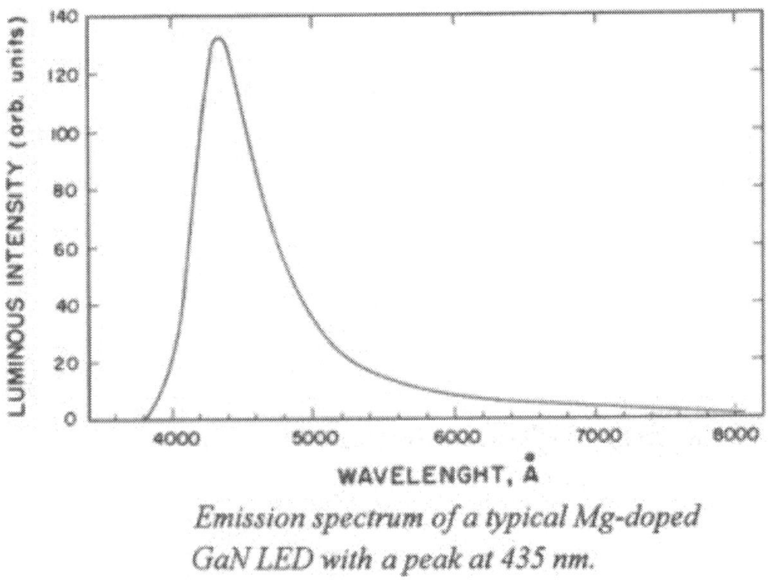

Figure 7. GaN:Mg LED emission spectrum [18].

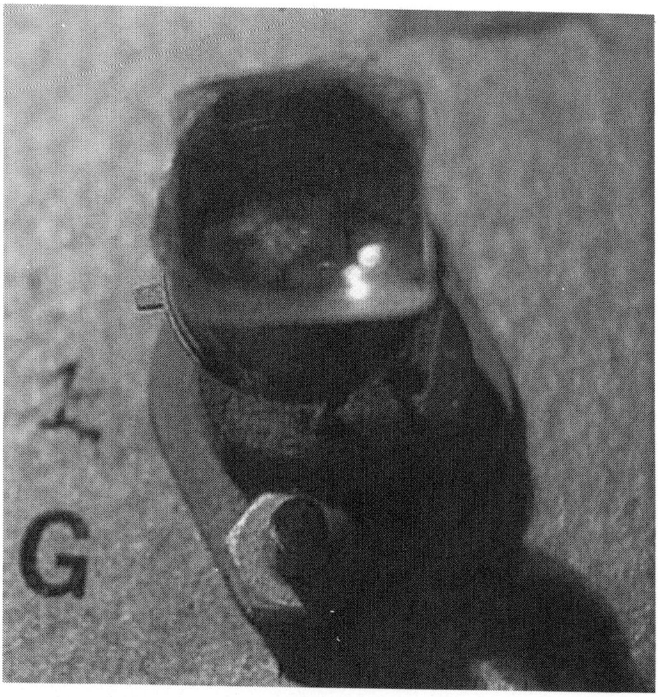

Figure. 8. Maruska's blue GaN LED still works in 2014.

Maruska finished his PhD thesis studies just in time for Christmas at the end of 1973. In January 1974 he returned to RCA Labs with a functioning blue LED in his attaché case, as he had promised. This device still works after all of these years. Its emission peak is at 425 nm. It can be seen at the Sarnoff Museum in The College of New Jersey near Trenton, NJ. A video demonstration of the device being connected to a power source and turned on appeared in the U.S. media in October 2014 and is available here: http://www.newsworks.org/index.php/thepulse/item/73871-the-blue-leds-nobel-worthy-lineage-traces-through-new-jersey.

RETURN TO RCA LABS

When Maruska returned to RCA in the fall of 1973, the department manager Dave Richman decided to pursue a new approach to nitride thin film growth. A novel method for preparing thin semiconductor films which did not involve the use of hydrogen chloride gas was just emerging. In 1971, Manasevit of North American Rockwell reported the first example of growing thin film nitride samples by the MOCVD (metal organic chemical vapor deposition) process which he had pioneered [19]. In this approach, gallium is transported into the growth zone as trimethyl-gallium, and no chlorine is used. Maruska quickly set about building one of the world's first MOCVD reactors, shown in Fig. 9.

Figure 9. Maruska's new MOCVD reactor, 1974.

WORLD LOSES INTEREST IN NITRIDES

Unfortunately, all of the RCA devices had relatively low efficiency and remained dim due to the heavy contamination always present in the HVPE process. It was clear that in 1974 these devices were far from being ready for commercial production, although Rhines and Maruska had received a patent for the Mg-doped blue GaN LED [20]. Due to the failure of RCA's computer business in 1971 (the company had to absorb a $250 million dollar loss) the company entered dire financial straits. Persistent difficulties with gallium nitride growth and loss of funding forced Tietjen to cancel the project early in 1974. In 1975, just a year after returning from Stanford with his doctorate, Maruska was laid off. Miller was let go as well and Pankove was reassigned to other projects within the lab (he then moved to the University of Colorado). This decision to close down the gallium nitride project was part of a major reduction in the size of the compound semiconductor program at RCA. The RCA team was so close, but they were not given the chance to cross the goal line.

Interestingly, firms that did not experience the same financial crisis as RCA, including many of its Japanese competitors, left gallium nitride research around the same time. Unable to make pn-junction LEDs, Bell Labs researchers were among the first to abandon the field. Other corporations followed suit, such as Philips, Hitachi, and Toshiba. A consensus grew among compound semiconductor researchers that gallium nitride was an intractable material and that working on it was a waste of time and resources.

The 1970s and early 1980s continued to be fruitless for GaN, although some research groups continued to fuss with the HVPE process, especially Guy Jacob with the French group working for Phillips [21]. In 1976, Jacob *et al.* used a standard Maruska-style HVPE reactor to grow GaN films and discussed their attempts to anneal Zn-doped films [22]. They found that when annealed in nitrogen or argon, GaN decomposes if $T > 970$ °C. When annealed in hydrogen it decomposes when $T > 600$ °C. However, when GaN films were annealed in steadily increasing ammonia pressures, then the net n-type carrier concentration was reduced. Other early researchers also attempted to anneal GaN films which contained the most common p-type dopants, Zn or Mg.

But breakthroughs in GaN device fabrication began starting in 1983. The introduction of the metal–organic chemical vapor deposition (MOCVD) process, pioneered by Manasevit at Rockwell in Los Angeles in 1971, allowed the introduction of cold reactor walls and is free of HCl, eliminating oxygen contamination from the growth vessel or from the HCl (which is always contaminated with water). New work on the growth of GaN was advanced by Asif Khan *et al.* at Honeywell starting in 1983 [23]. Kawabata *et al.* at Matsushita reported the first blue GaN LED grown by MOCVD in 1984 [24]. Using a rapid growth rate of 15 microns/h, they grew a 30-micron-thick n-type layer followed by a Zn-doped layer and got blue light emission.

RENEWED INTEREST AND MAJOR PROGRESS

Next, in 1986, Amano *et al.* applied the buffer layer concept, which greatly improved the structural quality of GaN films [25]. Efforts to reduce the densities of dislocations in $GaAs_{1-x}P_x$ films in the late 1960s had been centered on the growth of so-called buffer layers. A dislocation is actually either an extra plane of atoms or a missing plane, both of which cause stress in the lattice. It had been determined that by first growing a GaAs buffer layer, it is possible to eliminate some of the non-conforming lattice planes, and hence to reduce the dislocation density in the final $GaAs_{1-x}P_x$ device. With the GaN program, Miller had also sought to improve the lattice structure by growing films of undoped material to act as buffer layers, but the nitride program at RCA was terminated before he succeeded. Basically, the essential role of the buffer layer is to supply nucleation centers having the same crystalline orientation as the substrate and to promote lateral growth due to a decrease in interfacial free energy [26]. In 1994, Wickenden *et al.* demonstrated that GaN nucleation layers deposited at 540 °C on $\langle 0001 \rangle$ -oriented sapphire substrates have a measurable crystalline component, although the lack of absorbance features associated with the direct band gap of GaN suggests that the crystallite size is very small [27]. Upon annealing to higher temperatures, the crystallite size increases and the crystal perfection improves until at temperatures near those empirically determined to be optimum for epitaxial growth, it approaches that of good quality heteroepitaxial films. Most of the recrystallization of the nucleation layer occurs during the ramp from the deposition temperature to the final temperature.

What was really wrong with Maruska's HVPE method? The first attempt to explain the problem was given by Born and Robertson [28]. They determined that the electrical characteristics of the layers were affected by the major impurity in the system which was water. The commercial ammonia gas available at the time had a water content as high as

1000 ppm. At the same time, Monemar and his group [29] noted that the concentration of a large number of impurities in the material grown by Maruska's HVPE technique were at ppm levels. They suggested that an effort toward a refined growth technique might therefore give a material of considerably higher potential for applications such as blue-emitting LEDs than was presently available. An upgrading of the present growth procedure for GaN to avoid chemical contaminants at the ppm level seemed to be necessary before one could hope to prepare low-ohmic p-type GaN. Seifert indicated that oxygen was present in the form of H_2O contamination in NH_3, in the HCl gas, as well as a residual impurity in every hot-walled reactor system manufactured from silica. The last item refers to the reaction at high temperatures of HCl with SiO_2 which forms the walls of the reactor, to create $SiCl_4$ and release oxygen [30]. Fortunately, at that time, manufacturers of ammonia were cleaning up the contamination problems in this precursor gas as well. The undoped material now began to have much lower n-type carrier concentrations, between 10^{17} and 10^{18} cm^{-3}.

In 1988 Amano *et al.* reported on improved luminescence from Zn-doped GaN grown by MOCVD after electron beam annealing [31]. With Zn-doped films, they found a large increase in the intensity of 430 nm cathodoluminescence after electron beam irradiation in the scanning electron microscope. Contrast this with the RCA results in the 1970s which lacked any good results with the electron beam annealing of contaminated HVPE GaN:Mg material.

The big breakthrough came for Akasaki's group in 1989, when they announced low resistivity p-type samples with low energy electron beam (LEEBI) annealed Mg-doped GaN in a scanning electron microscope [32]. This process was explained by the following dynamic physical model firstproposed by Van Vechten *et al.* to explain GaN *p*-type conversion by LEEBI annealing. Electron beam excitation generates free electrons and holes which stimulate the breaking of acceptor-H bonds. Atomic H released from the dissociation of acceptor-H bond diffuses and may recombine with another atomic H to form H_2 molecules which can escape from the crystal [33].

The world's first nitride blue LED containing a pn junction made its debut at a conference in Los Angeles in 1989 [34]. The device was still not very bright, but the light it emitted was certainly blue. This blue gallium nitride LED apparently was based on carrier injection rather than impact ionization, although we must inquire why it provided blue rather than ultraviolet band gap emission.

The next step was to build a double heterostructure device, where the GaN active layer was sandwiched between two $Al_xGa_{1-x}N$ confinement layers. This took Akasaki and Amano another two years. Akasaki reported their new results at the fall meeting of the Materials Research Society in Boston in December, 1991 [35] (see Akasaki's results below in Fig. 10).

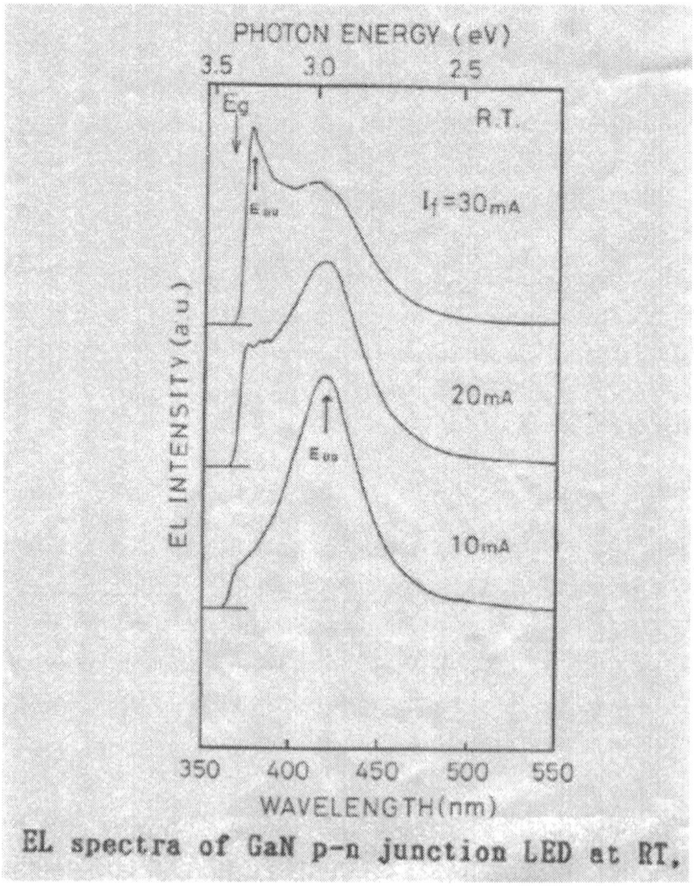

Figure 10. Akasaki's new brighter blue LED in 1991 [35].

Soon after Akasaki's important advances with buffer layers and low-energy electron beam annealing, Nakamura realized the advantage of thermal annealing of GaN films, which he grew by MOCVD, to produce highly conducting p-type layers [36]. Thermal annealing is a uniform process, unlike electron beam annealing. He thus was able to prepare bright and highly efficient blue LEDs based on GaN:Mg using standard thermal annealing [37]. See Fig. 11 below.

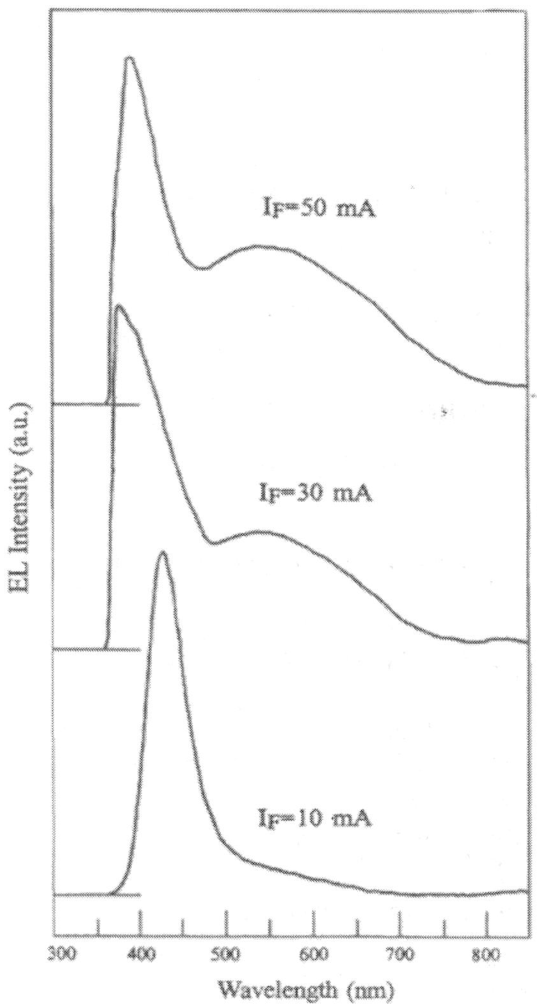

Emission spectra of the p-n junction GaN LED at different forward currents. Hole concentration of the p-layer of this sample was $1 \times 10^{17}/cm^3$.

Figure 11. Nakamura's blue emitting GaN:Mg LED [37].

BUT WHY WERE THEY BLUE?

The problem that needs to be addressed here is why are all of these GaN LEDs from the early 1990s were emitting blue light even though they are supposed to feature pn junctions? The band gap of GaN is 3.4 eV, so conventional band-to-band recombination should give emission centered at 364 nm, which is in the ultraviolet. All of the old results from the 1970s exhibited emission of blue light, usually in the range of 425–450 nm (seeFig. 12 below) [18]. An often invoked band diagram for GaN blue LEDs shows recombination across a band gap from the conduction band to the valence band. If this diagram actually applied to the case of GaN "blue" LEDs, then we should only see ultraviolet light emission. Obviously, this diagram is misleading. We would like to propose an alternative diagram.

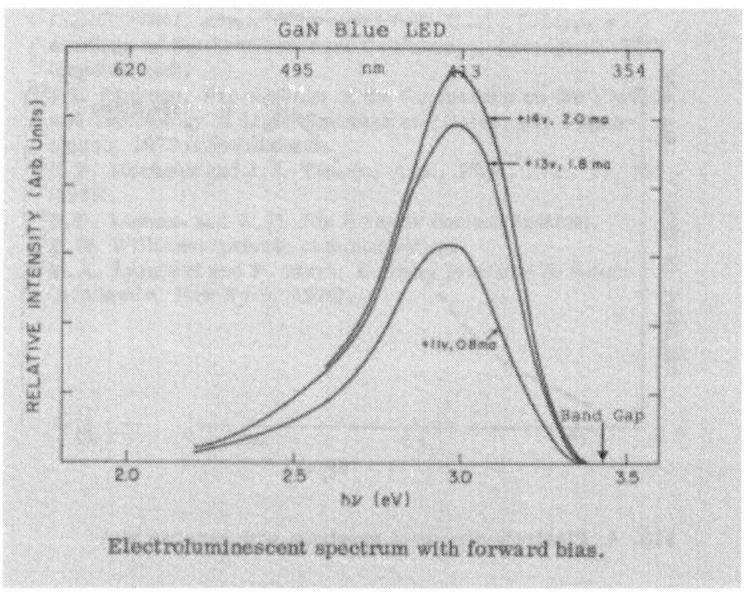

Figure 12. Electroluminescence spectrum of a Maruska LED in 1973 [18].

In 2015, when companies grow the p-type region of their commercial GaN LEDs, they continue to dope it with Mg. The doping concentration is about 1×10^{20} Mg/cm^3. If they make a Hall Effect measurement, they find about 1×10^{17} cm^{-3} holes after an *in situ*thermal anneal. Thus only 0.1% of the

Mg is activated. Then each unactivated Mg atom continues to have a hydrogen atom associated with it, and the H provides the third electron that satisfies its bonding requirements to replace a gallium atom in the crystal lattice [38]. Thus the present day commercial GaN LEDs will exhibit UV electroluminescence emitting 365 nm. But there is also blue electroluminescence at 450 nm. The blue light-emission requires traffic through the deep Mg–H levels. In order for the blue light-emission to occur, an electron must be removed from the deep levels by some process such as impact excitation. An electron that had been injected into the p-type material from the n-type region recombines with the newly created hole in the Mg–H complex, and then this electron completes its journey by recombining with a hole in the valence band, a hole that is associated with the bare Mg atom. This is shown in Fig. 13. This is the same mechanism for the blue emission that Stevenson and Maruska advanced in 1974 [39]. The Japanese researchers succeeded in adding the UV emission to their devices by activating 0.1% of the Mg atoms to provide holes in the valence band. That is why their devices still emitted blue light.

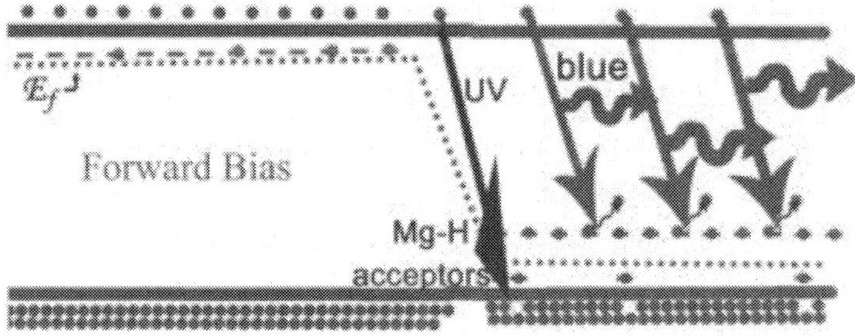

Figure 13. Correct band diagram when deep levels are present.

Let us summarize here the set of events which have been accumulating since 1971 which allowed researcher teams of Akasaki and Amano and Nakasaki to finally demonstrate a truly bright blue GaN-based LED in 1994. First of all, they used the MOCVD thin film growth technique from Manasevit of North American Rockwell, who first used it to deposit high quality nitride films in 1971. They based their early LED structures on the ones first demonstrated by Pankove with Zn doping, also in 1971. They used magnesium doping approach developed by Maruska and Rhines to acquire p-type doping, and their thermal annealing process to activate the Mg dopants. They also used the 1972 electron beam annealing process as an alternative approach to activate the Mg dopants. They also used Miller's buffer layer approach for reducing dislocation densities. And they were fortunate that much purer sources of ammonia became available over the years. But their breakthrough work was yet to come in 1994–1996.

TRUE BLUE NITRIDE LIGHT-EMITTING DEVICES

One cannot basically achieve a blue LED with a simple GaN film, because the band gap energy is 364 nm. To move the band gap energy into the visible spectrum, efforts were made to add InN to the GaN material to form an alloy with the band gap varying from 2.0 to 3.4 eV [40]. Nakamura and Mukai developed the first high quality alloy films of GaInN in 1992, which will lead to true blue and green LEDs and LDs featuring band-to-band recombination processes [41]. There is a very high vapor pressure of nitrogen over InN, making growth of nitride films containing significant concentrations of In quite difficult. But with alloys containing just a few percent indium, violet LEDs proved possible.

In 1993, Nakamura *et al.* described their early advanced p-GaN/n-InGaN/n-GaN DH violet LEDs. The InGaN films were grown by the two-flow metalorganic chemical vapor deposition (MOCVD) method. Sapphire with (0001) orientation (C face) was used as a substrate. Trimethylgallium (TMG), trimethylindium (TMI), monosilane (SiH_4), bis-cyclopentadienyl magnesium (CpzMg) and ammonia (NH_3,) were used as Ga, In, Si, Mg, and N sources, respectively. The thickness of the GaN buffer layer was approximately 25 nm. Next, the substrate temperature was elevated to 1020 °C, and the Si-doped n-GaN film was grown for 60 min to give a film 4 μm thick. Now the Si-doped active InGaN quantum well, with an approximate thickness of 10 nm, was deposited at 800 °C. After the Si-doped InGaN growth, the temperature was increased to 1020 °C to grow the Mg-doped p-type GaN film. Fig. 14a shows the electroluminescence (EL) of the InGaN/GaN DH LEDs at forward currents of 10 and 20 mA. The peak wavelength was 420 nm. Other peaks were not observed at all. They controlled the indium mole fraction (X) of InGaN active layer by changing the growth temperature or indium source flow rate during InGaN growth. Thus, they demonstrated that the peak wavelength of the EL emission of DH LEDs can be changed. This is shown in Fig. 14(b). This second group of DH LEDs was grown under the same conditions as the first batch, except for the Si-doped InGaN growth temperature, which was changed to 820 °C. The second batch of LEDs showed a peak wavelength is 411 nm. In view of these results, these violet emissions of InGaN/GaN DH LEDs can be assumed to take place through recombination between the electrons injected into the conduction band and holes injected into the valence band of the InGaN active layer [42].

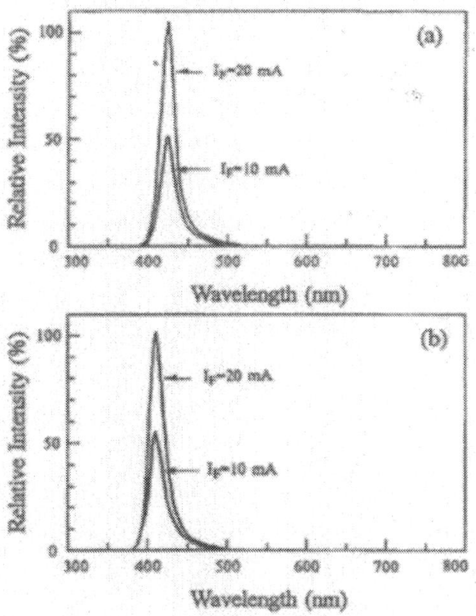

FIG. Electroluminescence spectra of the p-GaN/n-InGaN/n-GaN double-heterostructure blue LEDs which were grown under the same conditions except for the growth temperature of an InGaN active layer. The growth temperatures of InGaN were (a) 800 °C and (b) 820 °C.

Figure 14. Electroluminescence spectra of InGaN based LED devices [42].

Candela-class high-brightness InGaN/AlGaN double-heterostructure (DH) blue-light-emitting diodes (LEDs) with the luminous intensity over 1 cd were first fabricated in 1994. As an active layer, a Zn-doped InGaN layer was used for the DH LEDs. The typical output power was 1500 μW and the external quantum efficiency was as high as 2.7% at a forward current of 20 mA at room temperature. The peak wavelength of the electroluminescence was 450 nm. This value of luminous intensity was the highest ever reported for blue LEDs up to 1994 [43].

Akasaki and Amano reported LEDs based on GaInN in early 1995. A p-AlGaN:Mg/p-GaN:Mg/GaInN/n^+-GaN:Si/n-AlGaN:Si stacked heterostructure was fabricated. This heterostructure was grown on the n^+-GaN layer. A p-type stripe structure was prepared with the LEEBI treatment of the Mg-doped layers. The EL spectrum showed narrow and intense band-to-band transitions with the emission wavelength, determined by the In content of the GaInN layer, around 400 nm [44].

Finally, in 1995–1996, both of the research teams, Akasaki and Amano, and Nakamura, develop the critical device structure that made true blue emission possible from nitride semiconductors. The basis of their devices is an alloy of InN and GaN [45]. The formation of an $In_xGa_{1-x}N$ recombination section in the device allows them to exactly define the emission wavelength. They controlled the indium mole fraction (X) of InGaN active layer by changing the growth temperature or indium source flow rate during InGaN growth [46]. Thus, the peak wavelength of the EL emission of DH LEDs can be changed [47]. And by configuring it as a quantum well, it need not be doped. The quantum well acts like a "bucket", where holes are poured in from the p-type cladding layers, and electrons are injected from the n-type cladding layers on the other side. It does not matter if the Mg dopant in the p-type cladding layer provides any deep levels: the recombination is band-to-band in the well. And Nakamura's team reported the first actual blue InGaN quantum well laser diode device in January, 1996 [48]. There is just a single emission peak at 417 nm.

In November 1995, Akasaki and Amano reported the fabrication of a novel GaN/GaInN quantum well structure. This led to the first report of stimulated emission, by pulsed current injection at room temperature, from a group III nitride heterostructure using a very thin active layer. The nominally undoped GaN/GaInN MQW was fabricated by MOVPE on a sapphire (0001) substrate using an AlN buffer layer. The Multi-Quantum Well (MQW) structure was created on an unintentionally doped 1.8 μm thick n-GaN film. Each GaN barrier was 7.5 nm thick and each GaInN well was 2.5 nm thick. The InN molar fraction in the GaInN well was 0.07 [49]. A schematic diagram of the Akasaki/Amano device is shown in Fig. 15. Their first observation of stimulated emission from a quantum well nitride structure is shown in Fig. 16. In 1996, Akasaki and his group reported room temperature operation of a GaInN-based laser diode under current injection [50].

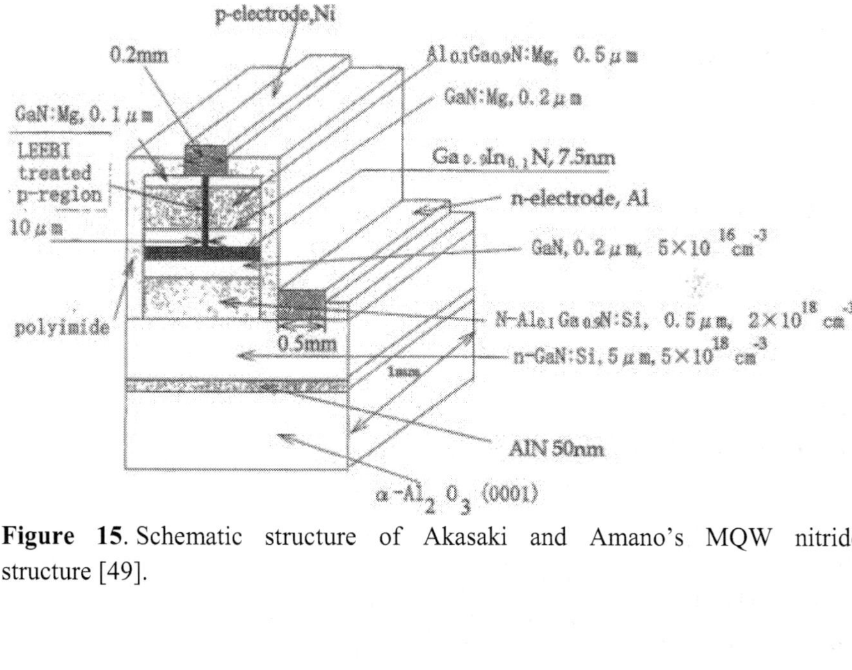

Figure 15. Schematic structure of Akasaki and Amano's MQW nitride structure [49].

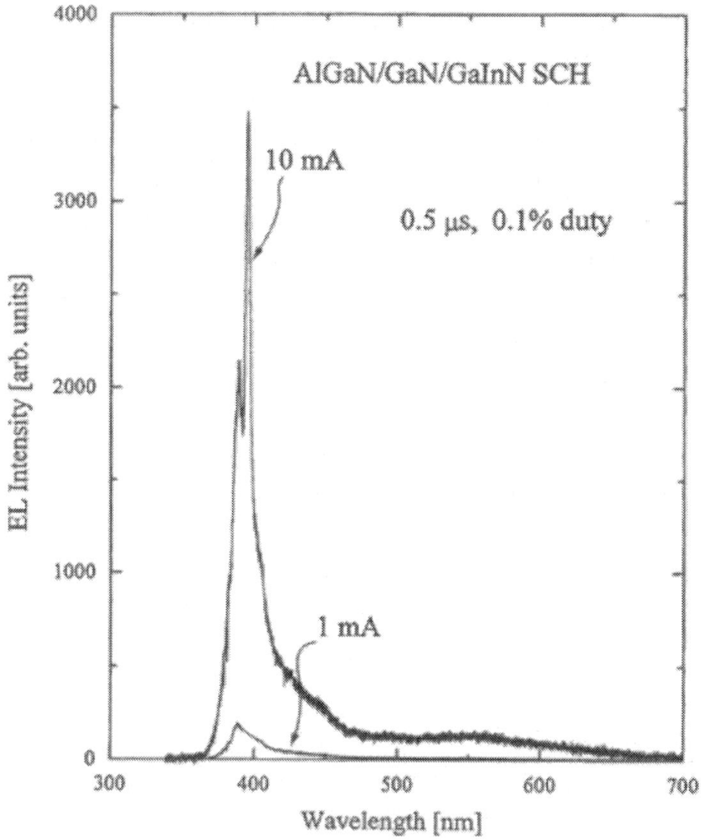

Figure 16. Akasaki and Amano's first quantum well nitride device [50].

Much work in this field has been performed by Nakamura and his team at Nichia [51]. The growth was performed by MOCVD, generally at atmospheric pressure. As usual, structures were grown on *c*-plane ⟨0001⟩ sapphire, with a low temperature (550 °C) GaN buffer, a thick n^+-GaN lower contact region, an n^+-InGaN strain-relief layer, an n^+-AlGaN cladding layer, a light-guiding region of GaN, then a multiquantum well region consisting of $In_{0.15}Ga_{0.85}N$ wells separated by $In_{0.02}Ga_{0.98}N$ barriers. The *p*-side of the device consisted of sequential layers of *p*-AlGaN, p^+-GaN light-guiding, *p*-$Al_{0.09}Ga_{0.92}N$ cladding and p^+-GaN contact. A ridge geometry was fabricated by dry etching down to the *p*-$Al_{0.08}Ga_{0.92}N$ layer), followed by formation of a mirror facet. The typical Nichia structure is shown in Fig. 17 [52]. The emission spectrum is shown in Fig. 18.

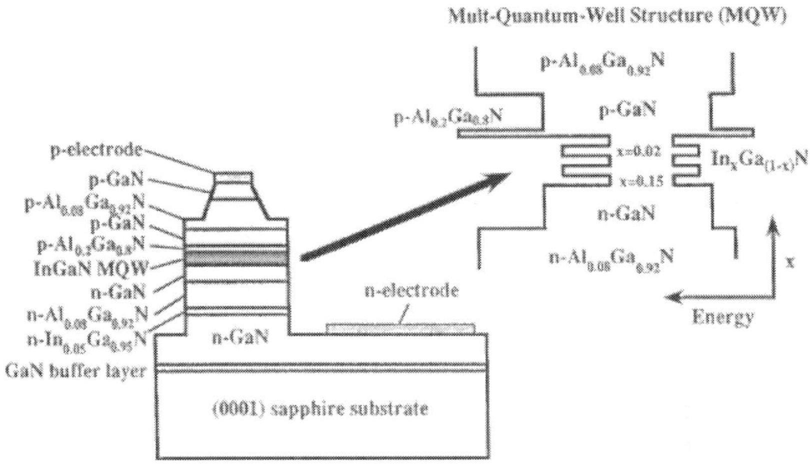

Figure 17. Schematic of the structure of a Nichia nitride laser structure [52].

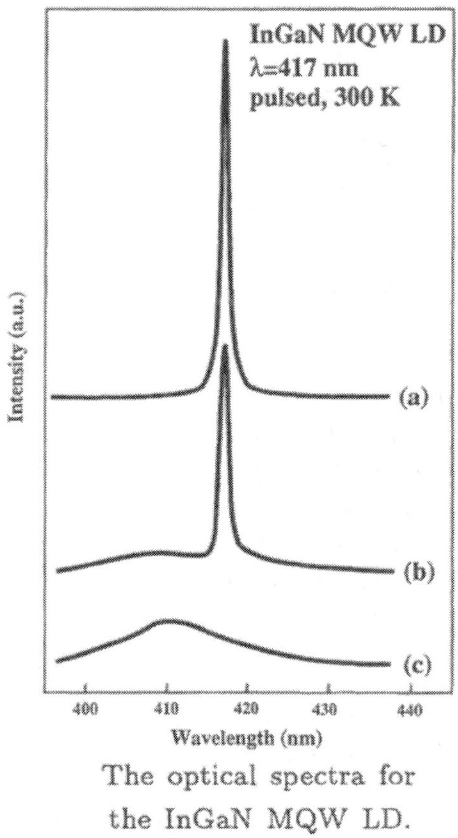

Figure 18. Emission spectrum of Nakamura's first true blue nitride laser diode [51].

These original quantum well devices warrant the 2014 Nobel Prize in Physics, which has been awarded to Akasaki, Amano, and Nakamura.

AFTERWORD

A few words about the subsequent careers of Rhines and Maruska. Wally Rhines graduated from Stanford University in 1972. He joined Texas Instruments and moved to Texas. He earned a Masters of Business Administration from Southern Methodist University. Rhines worked at Texas Instruments (TI) from 1972 to 1993, serving as executive vice president of the semiconductor group and president of the data systems group. Rhines became CEO of Mentor Graphics in 1993, when the company's annual revenue was about $340 million. The company passed $1 billion in revenue for the first time in 2011. Mentor Graphics is the largest private corporation in Oregon. He is no longer directly involved with nitride device studies, but maintains his interest in the field.

After being ejected from RCA in 1975, Herb Maruska worked at a number of companies, both large and small, in the semiconductor field. He was involved with both light-emitters and detectors (solar cells). In 1999 he joined a start-up company in Florida which sought to prepare free-standing GaN wafers for use as substrates for homoepitaxial growth of nitride devices (Fig. 19). He designed and built a new style combined MOCVD + HVPE reactor. Despite valiant efforts, there were no useful results. In 2003 he joined Asif Khan at the University of South Carolina where he worked on the MOCVD growth of wide-bandgap GaAlN devices. Maruska was forced to retire in 2006 due to health issues. He also maintains an interest in the nitride field.

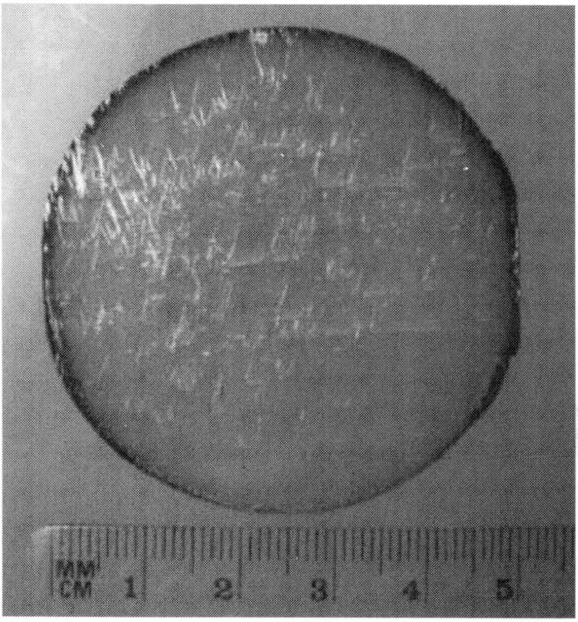

Figure 19. Free standing GaN wafer prepared by Maruska in 2003 in Florida.

REFERENCES

1. Potter RM, Blank JM, Addamiado A. J Appl Phys 1969;40:2253.
2. Violin EE, Kalnin AA, Pasynkov VV, Tairov DA, Yaskov YM. Silicon carbide – 1968. Pergamon Press; 1969. p. 231–41.
3. Edmond JA, King HS, Carter CH. Physica B 1993;185:453.
4. Pankove JI, Massoulie M. Bull Am Phys Soc 1962;7:88.
5. Holonyak Jr N, Bevacqua SF. Appl Phys Lett 1962;1:82.
6. Grimmeiss HG, Scholz H. Phys Lett 1964;8:233.
7. Logan RA, White HG, Wiegmann W. Appl Phys Lett 1968;13:139.
8. Tietjen JJ, Amick JA. J Electrochem Soc 1966;113:724.
9. Lorenz MR, Binkowski BB. J Electrochem Soc 1962;109:24.
10. Maruska HP, Tietjen JJ. Appl Phys Lett 1969;15:327.
11. Pankove JI, Miller EA, Richman D, Berkeyheiser JE. J Lumin 1971;4:63.
12. Pankove JI, Miller EA, Berkeyheiser JE. J Lumin 1972;5:84.
13. Pankove Jacques. Electroluminescent semiconductor device of GaN. U.S. Patent 3,683,240
1. filed 22.07.1971 and granted 08.08.1972..
14. Pankove JI, Miller EA, Berkeyheiser JE. J Lumin 1973;6:54.
15. Pankove JI, Miller EA, Berkeyheiser JE. RCA Rev 1971;32:383.

16. Pankove JI, Berkeyheiser JE, Miller EA. J Appl Phys 1974;45:1280.
17. Maruska HP, Rhines WC, Stevenson DA. Mater Res Bull 1972;7:777.
18. Maruska HP, Stevenson DA, Pankove JI. Appl Phys Lett 1973;22:303.
19. Manasevit HM, Erdmann FM, Simpson WI. J Electrochem Soc 1971;118:1864.
20. Stevenson David A, Rhines Walden C, Maruska Herbert P. Gallium nitride metal-semiconductor light-emitting diode; June 25, 1974
2. U.S. Patent 3,819,974..
21. Madar R, Jacob Guy M, Hallais Jean, Fruchart R. J Cryst Growth 1975;31.
22. Jacob G, Madar R, Hallais J. Mater Res Bull 1976;11:445.
23. Asif Khan M, Skogman RA, Schulze RG. Appl Phys Lett 1983;43:492.
24. Kawabata T, Matsuda T, Koike S. J Appl Phys 1984;56:2367.
25. Amano H, Sawaki N, Akasaki I, Toyoda Y. Appl Phys Lett 1986;48:353.
26. Akasaki I, Amano H, Koide Y, Hiramatsu K, Sawaki N. J Cryst Growth 1989;98:209.
27. Estes Wickenden A, Wickenden DK, Kistenmacher TJ. J Appl Phys 1994;75:5367.
28. Born PJ, Robertson DS. J Mater Sci 1980;15:3003.
29. Monemar B, Lagerstedt O, Gislason HP. J Appl Phys 1980;51:625.
30. Seifert W, Franzheld R, Butter E, Sobotta H, Riede V. Cryst Res Technol 1983;18:383.
31. Amano H, Akasaki I, Kozawa T, Hiramatsu K, Sawaki N, Ikeda K, et al. J Lumin 1988;40&41:121.
32. Amano H, Kito M, Hiramatsu K, Akasaki I. Jpn J Appl Phys 1989;28:L2112.
33. Van Vechten JA, Zook JD, Horning RD, Goldenberg B. Jpn J Appl Phys 1992;31:3662.
34. Amano H, Kitoh M, Hiramatsu K, Akasaki I. Inst Phys Conf Ser No 106: chapter 10 paper presented at lnt Symp GaAs and related compounds, Karuizawa, Japan; 1989.
35. Akasaki I, Amano H. Proc SPIE 1991;1361:138.
36. Nakamura S, Senoh M, Mukai T. Jpn J Appl Phys 1991;30:L1708.
37. Nakamura S, Mukai T, Senoh M. Jpn J Appl Phys 1991;30:L1998.
38. Nakamura S, Iwasa N, Senoh M, Mukai T. Jpn J Appl Phys 1992;31:1258.
39. Maruska HP, Stevenson DA. Solid State Electron 1974;17:1171.
40. Yoshimoto N, Matsuoka T, Sasaki T, Katsui A. J Appl Phys 1992;59:2251.
41. Nakamura S, Mukai T. Jpn J Appl Phys 1992;31:L1457.
42. Nakamura S, Senoh M, Mukai T. Appl Phys Lett 1993;62:2390.
43. Nakamura S, Mukai T, Senoh M. Appl Phys Lett 1994;64:1687.
44. Akasaki I, Amano H. J Cryst Growth 1995;146:455.
45. Nakamura S, Senoh M, Mukai T. Jpn J Appl Phys 1993;32:L8.
46. Nakamura S, Mukai T, Senoh M. Jpn J Appl Phys 1993;32:L16.

47. Nakamura S, Mukai T, Senoh M. J Appl Phys 1994;76:8189.
48. Nakamura S, Senoh M, Nagahama S, Iwasu I, Yamada T, Matsushita T, et al. Jpn J Appl Phys 1996;35:L74.
49. Akasaki I, Amano H, Sota S, Sakai H, Tanaka T, Koike M, et al. J Appl Phys 1995;34:L1517.
50. Akasaki I, Sota S, Sakai H, Tanaka T, Koike M, Amano H. Electron Lett 1996;32:1105.
51. Nakamura S, Senoh M, Nagahama S, Iwasa N, Yamada T, Matsushita T, et al. Appl Phys Lett 1996;68:2105.
52. Nakamura S, Senoh M, Nagahama S, Iwasa N, Yamada T, Matsushita T, et al. Jpn J Appl Phys, Part 2 1996;35:L217.

CITATION

Herbert Paul Maruska, Walden Clark Rhines, A modern perspective on the history of semiconductor nitride blue light sources, Solid-State Electronics, Volume 111, September 2015, Pages 32-41, ISSN 0038-1101, http://dx.doi.org/10.1016/j.sse.2015.04.010.

CHAPTER 6

PNP Pin Bipolar Phototransistors for High-Speed Applications Built in a 180 Nm CMOS Process

P. Kostov, W. Gaberl, M. Hofbauer, H. Zimmermann

Institute of Electrodynamics, Microwave and Circuit Engineering, Vienna University of Technology, Gusshausstr. 25/354, 1040 Vienna, Austria

ABSTRACT

This work reports on three speed optimized pnp bipolar phototransistors build in a standard 180 nm CMOS process using a special starting wafer. The starting wafer consists of a low doped p epitaxial layer on top of the p substrate. This low doped p epitaxial layer leads to a thick space-charge region between base and collector and thus to a high −3 dB bandwidth at low collector–emitter voltages. For a further increase of the bandwidth the presented phototransistors were designed with small emitter areas resulting in a small base-emitter capacitance. The three presented phototransistors were implemented in sizes of 40×40 μm^2 and 100×100 μm^2. Optical DC and AC measurements at 410 nm, 675 nm and 850 nm were done for phototransistor characterization. Due to the speed optimized design and the layer structure of the phototransistors, bandwidths up to 76.9 MHz and dynamic responsivities up to 2.89 A/W were achieved. Furthermore simulations of the electric field strength and space-charge regions were done.

HIGHLIGHTS

- Three speed optimized pnp phototransistors built in a 180 nm CMOS process are shown.
- A thick low doped intrinsic layer was implemented between base and collector.
- A thick base–collector space-charge region leads to high bandwidths.
- Optical characterisations were done at 410 nm, 675 nm and 850 nm.
- Bandwidths up to 76.9 MHz and dynamic responsivities up to 2.89 A/W were achieved.

INTRODUCTION

Photodiodes and phototransistors are the most commonly used photodetectors. By providing an additional gain, phototransistors are more suitable in some applications than photodiodes like for instance in low light scenarios. Nowadays photodetectors are built in silicon processes (these can also be standard CMOS or BiCMOS processes) or as III–V compound devices. Photodetectors integrated into standard silicon-based processes have several advantages over photodetectors realized in special silicon or III-V compound technologies. The main advantage of the implementation into a standard silicon process is the possibility for a cheap mass production of detector and circuitry together. The potential to combine the photodetector together with the readout circuitry to optoelectronic integrated circuits (OEICs) introduces further advantages of integrated solutions over wire-bonded solutions, due to the absence of bonding pads and bonding wires leading to less parasitic [1].

The physical property of silicon allows the material to be sensitive for wavelengths between 300 nm and 1100 nm. Light within this wavelength range will penetrate into the silicon and will be absorbed in it. Depending on the wavelength of the light, each photon can create one or more electron–hole pairs. A photon can create more than one electron–hole pair only if it is a high energy photon (corresponds to wavelengths below 375 nm) [2] and [3]. However, the photon can also be absorbed by free carriers without generating an electron–hole pair (free carrier absorption) [4]. Nevertheless, if silicon photodetectors are used in the visible and near-infrared range (400–900 nm), literature shows, that both mentioned effects can be neglected. Therefore for the visible light range the generation rate G of electron–hole pares in silicon can be simplified to [1]:

$$G(x,\lambda)=\Phi_0\alpha(\lambda)e^{-\alpha(\lambda)x} \quad (1)$$

The generation rate G is dependent on the depth x from the semiconductor surface, on the wavelength λ, on the photon flux Φ_0 of the incident light and on the absorption coefficient α. The absorption coefficient α decreases strongly for increasing wavelengthλ. To give two relevant examples: blue light (e.g. 430 nm) is absorbed in depths up to 0.2 µm while near infrared light (e.g. 850 nm) has a $1/e$ penetration depth of about 16.6 µm [1]. In a biased photodetector the electrical field separates the carriers. The wavelength dependent penetration depths lead to a wavelength dependent distribution of the total photocurrent into a drift and diffusion part, which further leads to wavelength dependent bandwidths and responsivities.

Photodiodes are the most common photodetectors. They can be implemented as a pn-junction or a pin-junction device. The difference between both structures is the additional low doped layer in the pin photodiode. In the junction region of the photodiode a space-charge region is formed. The space-charge region is thicker for the pin photodiode due to the additional low doped layer between the anode and cathode. For a vertical pin photodiode this will lead to a higher responsivity for deeper penetrating light and typically to a higher bandwidth since the drift current part is increased. PN photodiodes can be built in three different ways in a standard CMOS process using a p wafer: n-well/p-substrate, n+/p-substrate and p+/n-well. These diodes are characterized by a thin space-charge region and thus a small drift photocurrent, which leads to a relatively small bandwidth due to a large contribution of a diffusion current from photocharges generated below the space-charge region. However, by adding a thick intrinsic layer between the photodiode's anode and cathode, a thick space-charge region is formed and thus also a thick drift zone. A thicker drift zone leads to a larger fraction of drift current compared to diffusion current, which leads furthermore to higher bandwidths for low bias voltages. The advantage of higher bandwidths in pin over pn photodiodes makes them the most commonly used detector for high speed optical applications, for example in optical receivers [5]. Such a vertical pin photodiode with a −3 dB bandwidth of 2.8 GHz at $V_D = 20$ V and 0.2 A/W for 850 nm responsivity is presented in [6]. Another vertical pin photodiode with a responsivity of 0.135 A/W for 850 nm and a bandwidth of 1.6 GHz at $V_D = 15.5$ V is presented in [7] Furthermore pin photodiodes can also be implemented as lateral devices. In [7] lateral pin photodiodes with responsivities up to 0.15 A/W for 850 nm and bandwidths up to 3.1 GHz at $V_D = 15.5$ V are presented. For comparison the pn photodiodes presented in [8] reach a bandwidth of 80 MHz at $V_D = 8$ V and responsivities of 0.215 A/W and 0.15 A/W for 650 nm. As can be seen on the presented values, photodiodes can achieve only small responsivities in the range of 600–800 nm. This is due to a maximum quantum efficiency of $\eta = 1$. However, the maximum quantum efficiency cannot be reached in real devices due to reflection of the light at the surface as well as recombination of the photogenerated charges in the silicon.

Photodetectors with an inherent current amplification are introduced by avalanche photodiodes and phototransistors. Due to their inherent current amplification these photosensitive devices can improve the quantum efficiency limitation of photodiodes and reach values for η larger than 1.

Avalanche photodiodes achieve their inherent current amplification by an avalanche multiplication process. The drawback of this process is that voltages in the range of several tens of volts are needed to achieve sufficient electrical field strength for the inherent current amplification [9]. High voltages are a problem in integrated circuits and even more in modern low-voltage processes. However, new kinds of avalanche photodiodes, so called SPADs (single-photon avalanche photodiodes), can operate at smaller voltages than the standard avalanche photodiodes. In [10] two kinds of SPADs are presented with break-down voltages of 23.1 V and 16.1 V, respectively. The very narrow bias voltage range for linear operation of APDs and SPADs is a further disadvantage of these kinds of photodetectors.

Phototransistors have the big advantage over APDs that they do not need high voltages for their inherent current amplification. In BiCMOS, quite fast vertical npn phototransistors with responsivities of 2.7 A/W at 850 nm (due to the thin base–collector space-charge region) were reported [11]. Another BiCMOS SiGe phototransistor with a responsivity of 5.2 A/W at 850 nm is presented in [12]. In CMOS technology different types of phototransistors can be built, e.g. vertical bipolar phototransistors [13], [14], [15], [16], [17] and [18], lateral bipolar phototransistors [19], photo-MOSFETs [20], and so on. Vertical bipolar phototransistors with bandwidths up to 14 MHz and responsivities up to 98 A/W at 675 nm are presented in [13] and [14]. The CMOS lock-in amplifier presented in [15] uses pnp phototransistors with bandwidths of a few 100 kHz. Ref. [16] describes a pnp phototransistor with a bandwidth of 7.8 MHz at 638 nm. In [17] a photodiode and a npn phototransistor in a 65 nm CMOS process are presented. The phototransistor achieves a responsivity of 0.34 A/W and a bandwidth of 150 kHz. This work describes extended results of the phototransistors presented in [18].

WORKING PRINCIPLE

A vertical pnp phototransistor is formed in a CMOS process by implementing an n-well (base) in the p-substrate (collector) and a p^+-source/drain (emitter) in the n-well. Fig. 1a depicts the cross-section of a conventional pnp bipolar phototransistor. The pnp phototransistor can be used only in emitter-follower configuration since the substrate forms the collector. The phototransistor can be described as a photodiode (base–collector junction) and an internal bipolar junction transistor for the current amplification. Light with long wavelength penetrates deep into silicon and generates most charges in the base–collector space-charge region. There the charges are separated and electrons are swept into the base and holes into the collector. The electron accumulation in the base makes the potential of the base more negative, thus the forward bias voltage of the base-emitter diode increases and injection of holes from the emitter into the base becomes stronger. A large part of the injected holes can diffuse through the thin base and reach the base–collector space-charge region at the other end of the base area. This process leads to the amplification of the primary photocurrent. The inherent current amplification β is the relation between the collector current I_C and the primary photocurrent I_{PH}. Further literature on conventional CMOS phototransistors can be found in [1], [21] and [22].

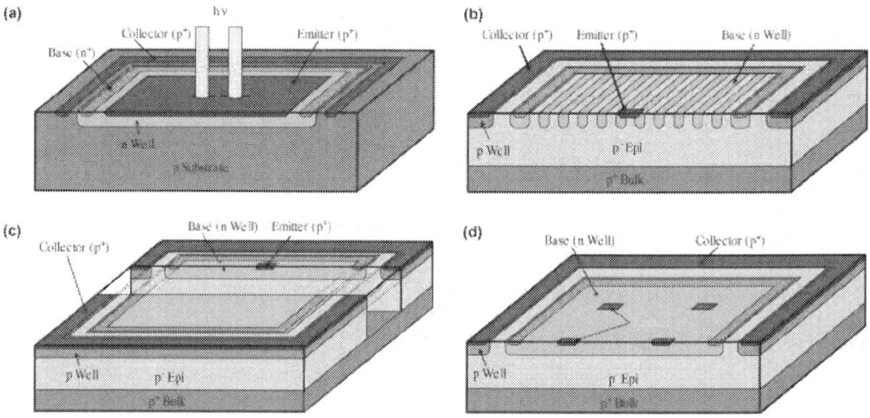

Figure 1. 3D depiction and cross section of the phototransistors: (a) common pnp phototransistor integrated in CMOS technology, PIN PNP phototransistors: (b) 50_BCenter$_E$, (c) 100_BEdge$_E$ and (d) 100_BQuad$_E$.

METHODOLOGY

We present three speed optimized pnp PIN phototransistors with sizes of 40×40 µm^{2}and 100×100 µm^2. The phototransistors are built in a 180 nm CMOS process without any process modification and have been fabricated by LFoundry. For achieving high bandwidths a special wafer was used. This wafer consists of a p^+ substrate with a low doped (5×10^{13} cm^{-3}) p epitaxial layer grown with a thickness of 15 µm on it. A thick base–collector space-charge region is formed therefore even at low voltages due to the present low doped p epitaxial layer. The PIN structure of the base–collector diode has the same properties like the mentioned PIN photodiodes in Section 1 and leads therefore to a fast separation of the photogenerated charges. As a consequence of the thick space-charge region the drift portion of the photocurrent for deep penetrating light rises and leads therefore to a higher responsivity compared to a conventional bipolar phototransistor like the one presented in [15].

The bandwidths and the responsivities of the phototransistors depend also on the layout properties of the base and emitter area. Eq. (2) shows the relationship between the −3 dB bandwidth of the phototransistor and the physical properties of the phototransistor [21]:

$$f_{-3\text{dB}} = \frac{1}{2\pi\beta \cdot \left(\tau_B + \frac{k_B T}{q I_E}(C_{\text{BE}} + C_{\text{BC}})\right)}, \tag{2}$$

where $f_{-3\text{dB}}$ is the −3 dB bandwidth of the phototransistor, β is the forward current gain of the phototransistor, τ_B is the base transit time, k_B is the Boltzmann constant, T is the absolute temperature, q is the elementary charge, I_E is the emitter current of the phototransistor, C_{BE} is the base-emitter capacitance and C_{BC} is the base–collector capacitance.

The −3 dB bandwidth depends on both space-charge region capacitances C_{BE} and C_{BC}, current gain β, base transit time τ_B and the emitter current I_E. In case of large capacitance values the −3 dB bandwidth is mainly dependent on both space-charge region capacitances C_{BE} and C_{BC}. For fast phototransistors a reduction of these capacitances is desired. In our presented phototransistors the capacitance C_{BC} is small compared to common phototransistors (e.g. Fig. 1a) due to the implemented thick low doped p epitaxial layer. Furthermore, a diminution of the capacitance C_{BE} is achieved by reducing the emitter area. Regarding the base doping two contrary effects have to be mentioned. On the one hand a reduction of the doping concentration in the base can lead to an increase of the bandwidth since both space-charge regions will reach deeper into the base leading to wider space-charge regions and thus smaller capacitances C_{BE} and C_{BC}. As a second consequence of larger space-charge region, the effective base width W becomes shorter leading to a reduced base transit time τ_B and thus also to an increased bandwidth. On the other hand, thicker space-charge regions will cause a reduction of the electric field strength inside the space-charge regions for the same potential conditions, leading to a slower device since the velocity of the charge carriers in the space-charge region will be reduced. This can be seen from the simulations in Section 5.1. Furthermore the reduction of the base doping concentration leads to a higher inherent current gain $\beta$and thus to a higher collector and emitter current I_C and I_E. The higher emitter current will cause according Eq. (2) a higher bandwidth. Hence it follows that for phototransistors with generally small junction capacitances (e.g. optimized emitter layout, PIN structure) the frequency behavior of the device becomes more dependent on the other parameters like β, τ_B and I_E. A further increase of the bandwidth can be achieved by reducing the existing perimeter capacitance and further by reducing the size of the whole phototransistor. A reduction of the perimeter capacitance was achieved by adding a 3 μm gap with low doping between the n-well base and the p-well collector contact. However, the bandwidth of the PIN phototransistor is lower than the bandwidth of the PIN photodiode due to the both junction capacitances and the base transit time τ_B which is no issue in the photodiode.

As mentioned above, the bandwidth will increase when reducing the emitter area. On the other side, a smaller emitter area results in a smaller responsivity if the whole device is illuminated. The reason for the smaller responsivity is the higher charge recombination probability due to a longer travel distance. Charges which are not generated directly under the emitter have to travel longer distances through the base to reach the base-emitter junction. These charges will contribute to the photocurrent only if their lifetimes are longer than the time they need to reach the base-emitter space-charge region. Regarding the responsivity the base doping concentration plays also a crucial role. For a smaller base doping concentration a thinner effective base width W is formed leading to a decreased travel distance of the generated charges through the base. This increases the probability for the charges to reach the base-emitter junction and thus also increases the inherent current amplification β. The inherent current amplification β is proportional to the emitter doping concentration N_E and the Gummel number N_G, which describes the doping concentration N_B of the base [22]:

$$N_G = \int_0^W N_B(x)dx, \tag{3}$$

$$\beta \propto \frac{N_E}{N_G} \tag{4}$$

As a consequence we can say that for achieving high responsivities the phototransistors should be designed with a large emitter over the whole photosensitive area together with a low doped base. Whereas for achieving higher bandwidths the phototransistors should be designed with very small emitter areas together with tendencially higher doped bases. It should be furthermore mentioned that by reducing the doping concentration of the base the probability for reach-through between collector and emitter increases, which should be avoided. However, the presented phototransistors are designed for high bandwidth applications and are implemented thus with small emitter areas. In the measurement section of this paper it can be seen that no reach-through arises for even high collector–emitter voltages.

IMPLEMENTED PHOTOTRANSISTORS

The three realized phototransistor versions (Fig. 1b–d) were built in $40 \times 40\ \mu m^2$ and $100 \times 100\ \mu m^2$ and have due to different layout designs of base and emitter different characteristics:

- *50_BCenter$_E$*: This phototransistor was designed with a striped base. These n-well stripes have a width of 0.5 µm and are separated by 0.5 µm wide gaps between them. During the fabrication the n-well stripes will diffuse due to the thermal budget into one single layer with the half doping concentration of a base consisting of a full n-well. This is also the reason why this device is called 50_BCenter$_E$. The emitter of this device has the size of $0.74 \times 0.74\ \mu m^2$ and is placed in the center of the photosensitive area.

- *100_BEdge$_E$*: As can be seen from the name, this device consists of a full n-well base and an emitter at the edge of the photosensitive area. It has a slightly larger emitter area compared to the 50_BCenter$_E$ phototransistor due to the demands of the design rule specifications. The emitter area has a size of $2.18 \times 0.32\ \mu m^2$ and is formed by a p^+ drain/source implant. The idea for having the emitter at the edge of the photosensitive area is based on the idea of implementing an anti-reflection layer and an optical window etch on top of the photosensitive area and thus increasing the responsivity of the device. However, for a better comparison devices without an optical window edge were used.
- *100_BQuad$_E$*: This device has a similar layout as the device presented before. The difference between both phototransistors is that this device has four separate emitter areas. Each emitter area is placed in the center of each quadrant of the photosensitive area. All emitter areas are connected with minimum width metal lines on top of the phototransistor.

RESULTS AND DISCUSSION

We characterized the presented phototransistors by optical DC and AC measurements. Optical DC measurements were done using a laser with 850 nm wavelength for the characterization of the phototransistors output characteristic and DC responsivity. Also, the spectral responsivity over the whole visible light spectrum was measured. Dynamic responsivity and bandwidth measurements of the phototransistors were done at 410 nm, 675 nm and 850 nm, respectively. The following equipment was used for the characterization of the phototransistors: the three mentioned laser sources, a monochromator for measuring the spectral responsivity, an optical attenuator and optical power meter for monitoring the light power, source-meter-units (SMUs) for applying voltages and measuring current, an oscilloscope for measuring the AC responsivity and a vector network analyzer for measuring the frequency step response, respectively. All optical paths were calibrated with a fast optical reference photodiode. Furthermore the electric field strength and the space-charge regions of the phototransistors were simulated.

Electric field strength and space-charge region simulation

Electric field strength and space-charge region simulations were done in order to make clear the variations between different collector–emitter voltages V_{CE} and different base doping concentrations. The phototransistor 50_BCenter_E was simulated in dark light conditions at $V_{CE} = -2$ V, -5 V and -10 V always with floating base. Fig. 2 depicts the electric field strength in this phototransistor. In this figure the thick drift zone of the base–collector space-charge region and the strength of the electric field are noticeable. The peaks close to $Y = 0$ µm are due to the contacts of collector, base and emitter of the phototransistor. For a better comparison of the electric field strength in the base–collector space-charge region for the three collector–emitter voltages the electric field strength was limited in the plot. Strong electric field strength peaks in the contact region are not shown to improve the scaling of the overall picture. The electric field strength in the base-emitter space-charge region reaches 35 kV/cm. The borders of the space-charge regions for the 50_BCenter_E phototransistor are depicted in Fig. 3. The phototransistor was simulated with a single n-well layer with the half doping concentration since the n-well stripes of this device will diffuse during the production into a single layer. Recognizable is the difference in the thickness of the base–collector space-charge region due to different collector–emitter voltages. In Fig. 4 the difference of the electric field between the 50_BCenter_E and 100_BEdge_E phototransistor is depicted. Due to the lower doping concentration inside the base it is apparent that the 50_BCenter_Ephototransistor has wider space-charge regions but lower electric field strength compared to the 100_BEdge_E phototransistor. Thinner space-charge regions and thus higher electric field strengths in them will lead to a faster drift component. This can be seen in the bandwidth results section (Section 5.3.2).

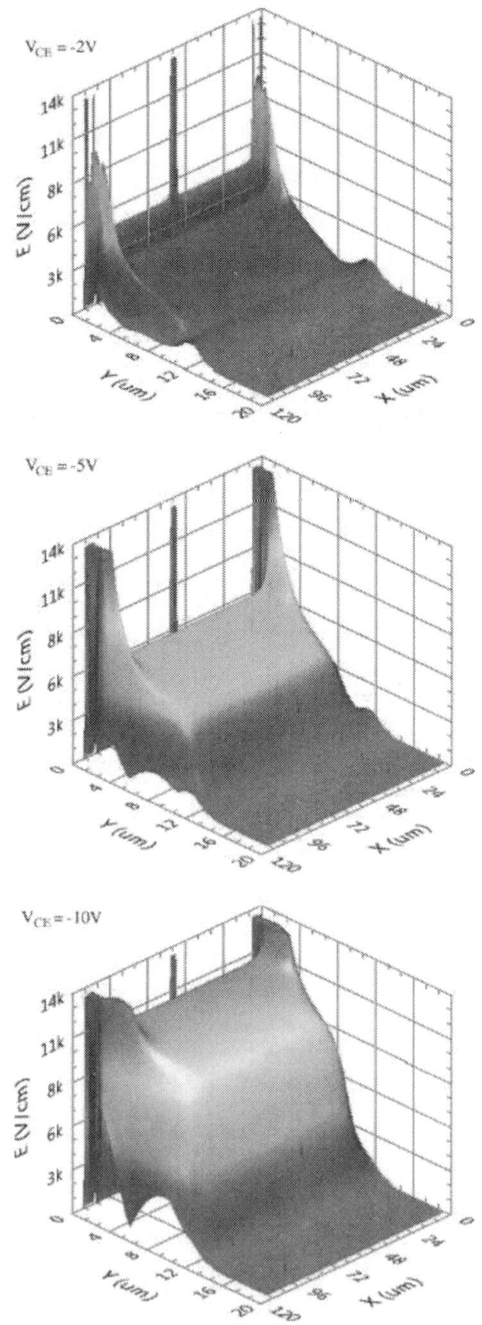

Figure 2. Simulated electric field strength for the $50_{B}Center_{E}$ phototransistor at three different V_{CE} voltages.

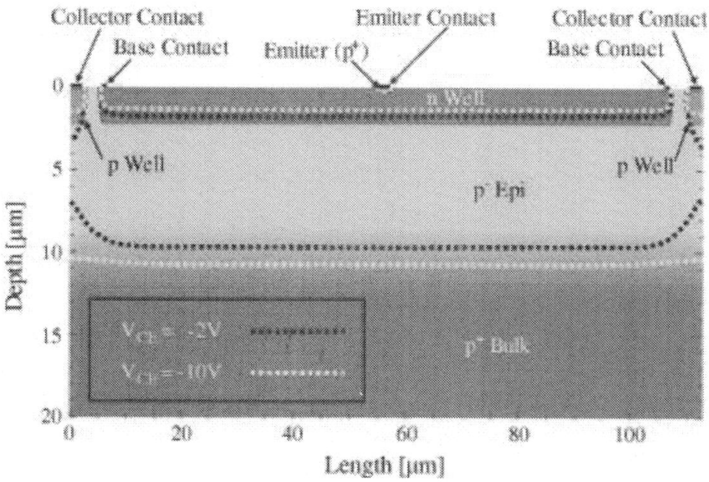

Figure 3. Borders of the space-charge region at $V_{CE} = -2$ V and -10 V for the 50_BCenter_E phototransistor.

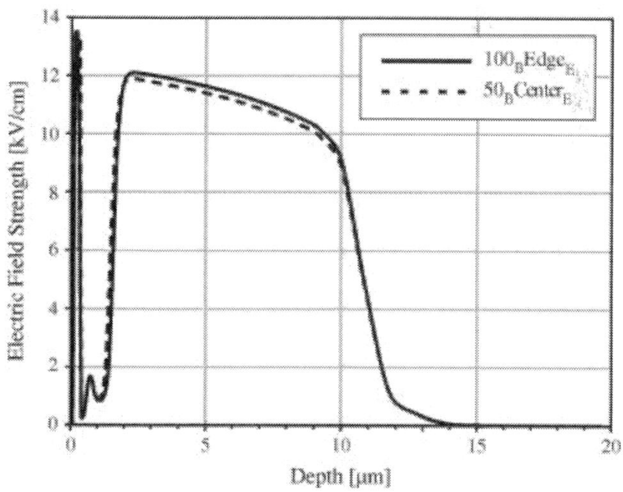

Figure 4. Cross-section of the electric field strength at the center of the emitter of the 50_BCenter_E and of the 100_BEdge_E phototransistor.

DC characterization

The DC characterization was split in two different measurement setups. First, the output characteristics were characterized and secondly the spectral responsivity was measured.

I_C vs. V_{CE} curve family and responsivity at 850 nm

For the characterization of the I_C vs. V_{CE} curve family (output characteristics) a light source with 850 nm wavelength was used. The operating point was changed by sweeping the collector–emitter voltage V_{CE} from 0 V to −13 V and the optical power between −37.7 dBm and −8.3 dBm. Thereby the light power was changed by an optical attenuator and monitored via an optical power meter. A source-meter-unit (SMU) was used on one hand to change the collector–emitter voltage and on the other hand to measure the collector current, respectively. The base contact was left floating. Fig. 5depicts the output characteristics of the 100 × 100 μm^2 50$_B$Center$_E$ phototransistor. In this figure, it can be seen that for voltages V_{CE} up to −13 V no reach-through occurs. All other devices show an almost similar output characteristic like the depicted one.

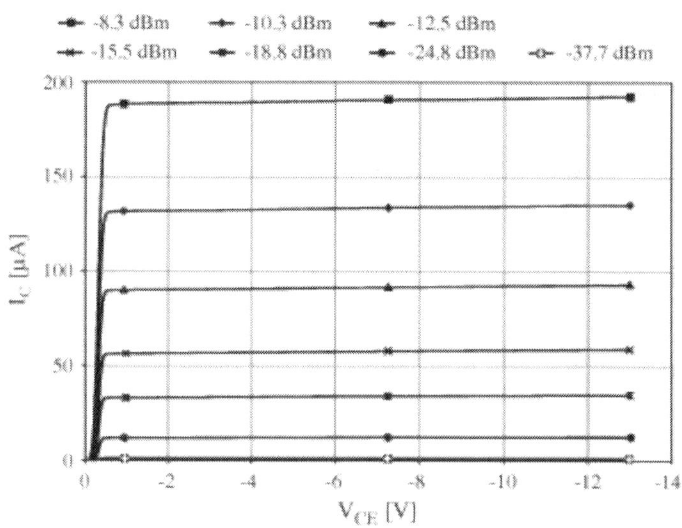

Figure 5. Output characteristics of the 100 × 100 μm^2 50$_B$Center$_E$ phototransistor at 850 nm for different optical power with floating base.

The calculated DC responsivity for the 50$_B$Center$_E$ and the 100$_B$Edge$_E$ 100 × 100 μm^{2}phototransistors at V_{CE} = −10 V is depicted in Fig. 6. The responsivity decreases for increasing optical light power due to a reduced gain in the phototransistor, which is caused by a change in the operating point as described in [13]. Thereby the higher optical light power induces a high collector current I_C which leads to a current density being larger than the critical current density $j_c \sim N_C$ (N_C is the collector doping concentration) given by the Kirk effect [23] and [24] (also called base push out effect). Due to the doping concentration of 5×10^{13} cm^{-3} in the collector layer (this is the thick p-epitaxial layer) instead of usual collector doping levels above 10^{15} cm^{-3}, the critical current density is reduced by about two orders of magnitude. Therefore, even at rather low collector currents, the current gain β reduces with increasing optical input power.

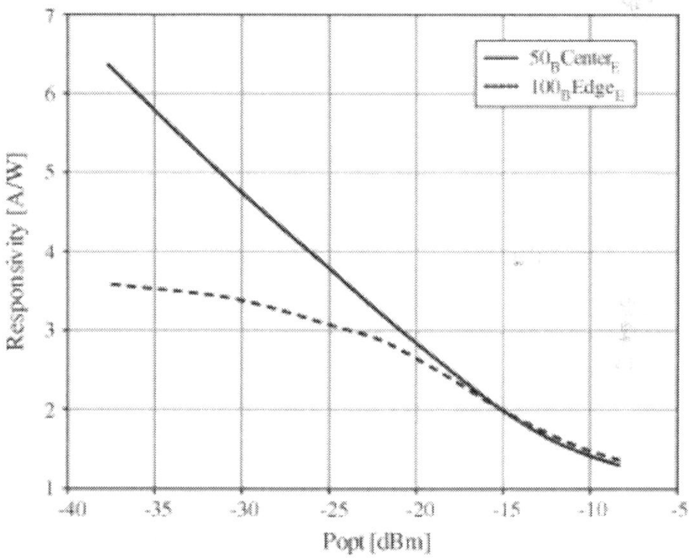

Figure 6. DC responsivities of the 100×100 μm^2 50_BCenter$_E$ and 100_BEdge$_E$ phototransistors at 850 nm at $V_{CE} = -10$ V.

At weak optical light power the 50_BCenter$_E$ phototransistor shows a higher gain due to the lower doped base and thus a higher inherent current gain β. Phototransistor 100_BQuad$_E$ has due to more emitter area a little bit higher responsivity as the 100_BEdge$_E$phototransistor. For all phototransistors the responsivity does not change much for different V_{CE}, since the collector current I_C is less dependent on different V_{CE} in the forward active region (see Fig. 5). In Table 1 the DC responsivities for the three

100 × 100 μm^2 phototransistors at different V_{CE} and different optical light power are shown.

Table 1. DC responsivity in A/W for the three 100 × 100 μm^2 phototransistors for three different collector–emitter voltages at 850 nm and an optical power of −37.7 dBm and −15.5 dBm.

	$P_{opt} = -37.7$ dBm			$P_{opt} = -15.5$ dBm		
	$V_{CE} = -2$ V	$V_{CE} = -5$ V	$V_{CE} = -10$ V	$V_{CE} = -2$ V	$V_{CE} = -5$ V	$V_{CE} = -10$ V
50_BCenter$_E$	6.12	6.23	6.37	2.01	2.03	2.07
100_BEdge$_E$	3.5	3.51	3.52	2	2.02	2.05
100_BQuad$_E$	3.57	3.57	3.59	2.63	2.65	2.68

Spectral responsivity

The spectral responsivities of the phototransistors were measured by means of a monochromator. It was used to sweep the wavelength of the light from 400 nm to 900 nm. The optical light power of the monochromator changes thereby between −35.7 dBm and −26 dBm as depicted in Fig. 7a. The optical output power of the monochromator with attached optical fiber was measured with a calibrated reference photodiode. Then the fiber was adjusted to the phototransistors, whereby all light fell into their light sensitive area. A SMU was used to set the collector–emitter voltage and to measure the emitter current. The responsivity was then calculated from the measured emitter current and this incident light power. Fig. 7b shows the responsivity of the 50_BCenter$_E$ and the 100_BEdge$_E$ 100 × 100 μm^2 phototransistors at $V_{CE} = -2$ V. The phototransistor 50_BCenter$_E$ shows a higher responsivity compared with the 100_BEdge$_E$ phototransistor (compare with Fig. 6). A maximum responsivity is measured for all phototransistors in the red wavelength range. The oscillations, which can be seen in the spectral responsivity, are due to the influence of optical interference in the full oxide and passivation stack. They can be avoided by applying an optical window etch step together with an antireflection coating on top of the photosensitive area.

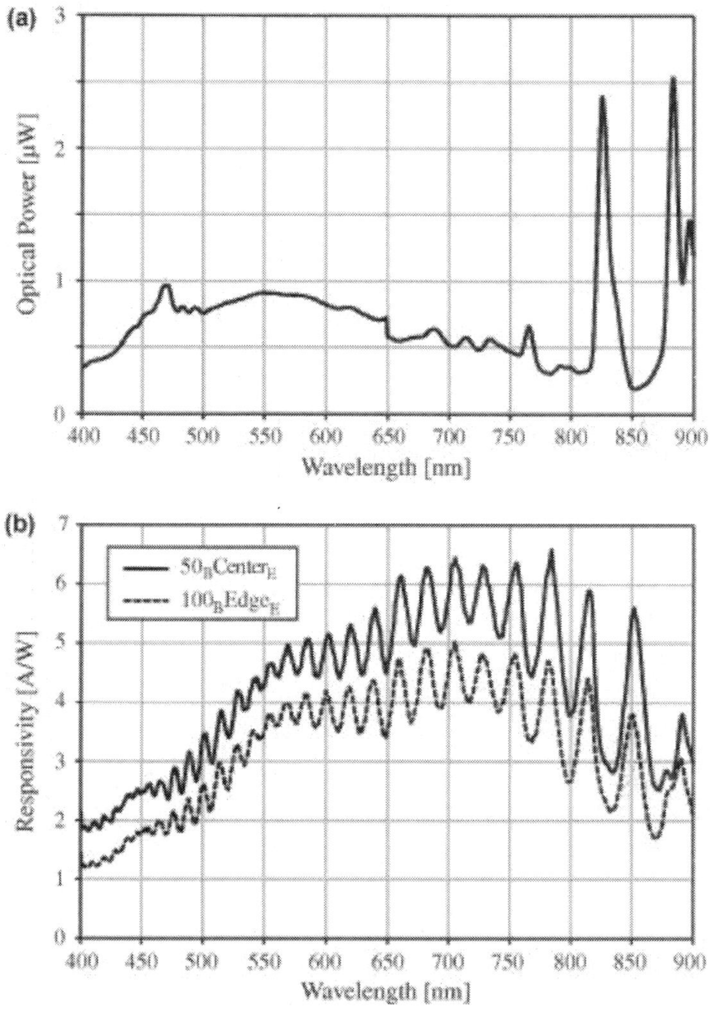

Figure 7. Spectral measurements: (a) emitted optical power of the monochromator and (b) spectral responsivity of the 100 × 100 µm² 50_BCenter$_E$ and 100_BEdge$_E$ phototransistors at $V_{CE} = -2$ V.

AC characterization: responsivity and bandwidth

AC responsivity measurements of the phototransistors were done at three different wavelengths: 410 nm, 675 nm and 850 nm. By using the same optical light power for the mentioned wavelengths different collector currents I_C will arise on the one hand due to different responsivities at each wavelength and on the other hand on different energy per photon for different light color. Therefore a comparison of the individual wavelengths would not be absolutely correct since the phototransistor would operate in various operating points for different wavelengths. For a better comparison, an alignment of the optical light power was applied to meet the same collector current I_C. Thus the mean optical light power at 410 nm was set to −12.7 dBm, at 675 nm it was set to −19.2 dBm and at 850 nm it was set to −15.8 dBm. Due to the used laser sources the extinction ratio was 2.00, 2.74 and 1.48 for 410 nm, 675 nm and 850 nm, respectively. During the AC characterization also different operating points were used for the phototransistors. Additionally the collector–emitter voltage V_{CE} was set to −2 V, −5 V and −10 V and the base current I_B was varied from floating condition (0 μA) to 1 μA, 2 μA, 5 μA and 10 μA. A bias-tee element together with an on-chip base-resistor was used to set the different operating points.

AC responsivity

The dynamic responsivity was measured for the three mentioned wavelengths at a frequency of 630 kHz. The phototransistors were connected in emitter follower configuration and their output signal was capacitively coupled to the oscilloscope via a bias tee element. All phototransistors achieve rather small responsivities mainly due to the small emitter areas. These small emitter areas implicate a higher recombination probability of the charges inside the base area. This is caused by the fact that the charges have to travel longer distances to reach the emitter area. Only charges which are generated directly under the emitter have to pass only a short distance through the base to reach the base-emitter space-charge region. However, the phototransistors presented here are designed for achieving high bandwidths. In Table 2 the dynamic responsivities for the different phototransistors at different collector–emitter voltages V_{CE}, floating base and different wavelengths are presented. The highest responsivity of 2.89 A/W is achieved for the 100_BQuad$_E$ phototransistor at $V_{CE} = -10$ V and 675 nm. By applying the above mentioned base currents the responsivity slightly decreases. This is caused by an arising base-push out effect and a reduced current gain β due to high base currents [23] and [24]. Due to the demands of the design rule specifications the 100_BEdge$_E$phototransistor has a slightly larger emitter area which results also in a slightly higher responsivity compared to the 50_BCenter_E phototransistor.

Table 2. Dynamic responsivity in A/W for the three $100 \times 100\ \mu m^2$ phototransistors for two different collector–emitter voltages at 410 nm, 675 nm, 850 nm and floating base.

	$\lambda = 410$ nm		$\lambda = 675$ nm		$\lambda = 850$ nm	
	$V_{CE} = -2$ V	$V_{CE} = -10$ V	$V_{CE} = -2$ V	$V_{CE} = -10$ V	$V_{CE} = -2$ V	$V_{CE} = -10$ V
50_BCenter_E	0.45	0.48	1.93	1.95	1.34	1.36
100_BEdge_E	0.47	0.51	2.06	2.08	1.4	1.5
100_BQuad_E	0.71	0.74	2.81	2.89	2.2	2.34

Bandwidth

The bandwidth characterization of the phototransistors was done by the means of a vector network analyzer (VNA). The phototransistors 50_BCenter_E and 100_BEdge_E show nearly the same bandwidths. Phototransistor 100_BQuad_E achieves lower bandwidths due to more emitter area and thus a larger base-emitter capacitance C_{BE}. For the 50_BCenter_E phototransistor the high bandwidth is mainly caused by the thin effective base width and thus short base transit time and furthermore smaller junction capacitances C_{BC} and C_{BE}. Regarding the 100_BEdge_E phototransistor, which is even slightly faster, the high bandwidth is dominated by the slightly higher electric field strength in the space-charge regions. These space-charge regions are slightly smaller due to the higher doped base area. Furthermore the small sized phototransistors show a higher bandwidth compared to the large sized ones. Here the smaller base–collector capacitance and a smaller perimeter capacitance are the main reasons. A higher bandwidth can be achieved by increasing the collector–emitter voltage V_{CE} and also by applying a base current I_B. The higher V_{CE} leads to wider space-charge regions, causing smaller junction capacitances C_{BE} and C_{BC}, and consecutively to a thinner effective base width, causing a shorter base transit time. Furthermore the higher V_{CE} causes a stronger electric field strength inside the device (see Fig. 2). Fig. 8 shows the frequency response dependency on the size of the phototransistors as well as on the collector–emitter voltage V_{CE} at

850 nm for the $100_B Edge_E$ phototransistor. The −3 dB bandwidths are 12.0 MHz and 25.7 MHz for the 100 × 100 μm² sized phototransistors and 14.2 MHz and 50.7 MHz for the 40 × 40 μm² sized phototransistors at V_{CE} = −2 V and −10 V, respectively.

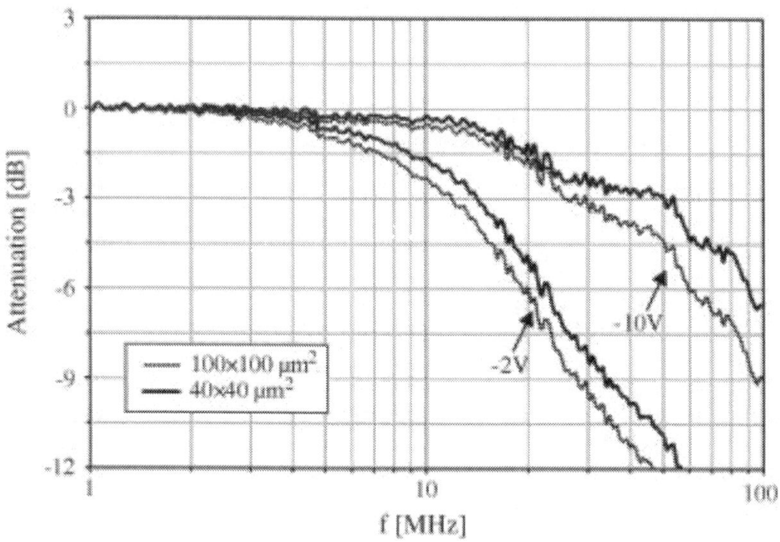

Figure 8. Frequency response of the 40 × 40 μm² and 100 × 100 μm² $100_B Edge_E$ at 850 nm and floating base.

In Fig. 9 the bandwidths for the 100 × 100 μm² $100_B Edge_E$ phototransistor at 850 nm and different operating points are depicted. The bandwidth increases with the base current until the collector current density reaches a maximum. At this point the largest homogeneous electric field exists in the base–collector space-charge region. By driving a higher collector current the charges cannot be carried completely by the electric field anymore [22]. Beyond the maxima in Fig. 9 the base push-out effect arises and leads to a spreading of the effective base into the collector [23] and [24]. Thus the effective base width gets wider and the base transit time increases, leading according to Eq. (2) to a reduced bandwidth. However, it should be mentioned that the position of the maxima inFig. 9 depend on the collector–emitter voltage V_{CE}. Hence an increase of V_{CE} leads also to an increase of the corresponding maximum collector current density and thus furthermore to an increase of the base current for the bandwidth maximum.

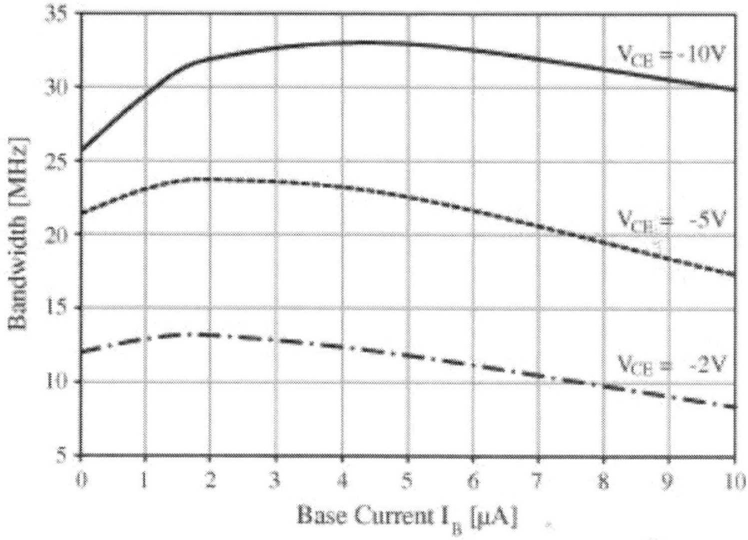

Figure 9. Bandwidth dependence on base current at three different V_{CE} voltages for the 100 × 100 μm² $100_B Edge_E$ at 850 nm.

The difference in the bandwidth at different wavelengths for the 40 × 40 μm² $100_B Edge_E$ phototransistor at $V_{CE} = -10$ V is shown in Fig. 10. The −3 dB bandwidths are 50.7 MHz at 850 nm, 76.9 MHz at 675 nm and 60.5 MHz at 410 nm for this phototransistor. In Table 3 the bandwidths for the three presented phototransistors at $V_{CE} = -2$ V and −10 V and floating base at the three different wavelengths are shown. The upper part of the table presents the results for the 40 × 40 μm² and the lower one for the 100 × 100 μm² devices. Noticeable is that phototransistor $100_B Quad_E$ shows a higher bandwidth for $V_{CE} = -2$ V compared to both other phototransistors. This is caused due to a shorter diffusion distance for the generated charges. Phototransistor $100_B Edge_E$ achieves the highest bandwidth due to stronger electric field strength in the space-charge regions (see Fig. 4). However, all devices achieve a maximal bandwidth at 675 nm due to an optimal light penetration depth. Thereby the main part of the charges is generated in the thick base–collector space-charge region and thus directly in the electric field zone.

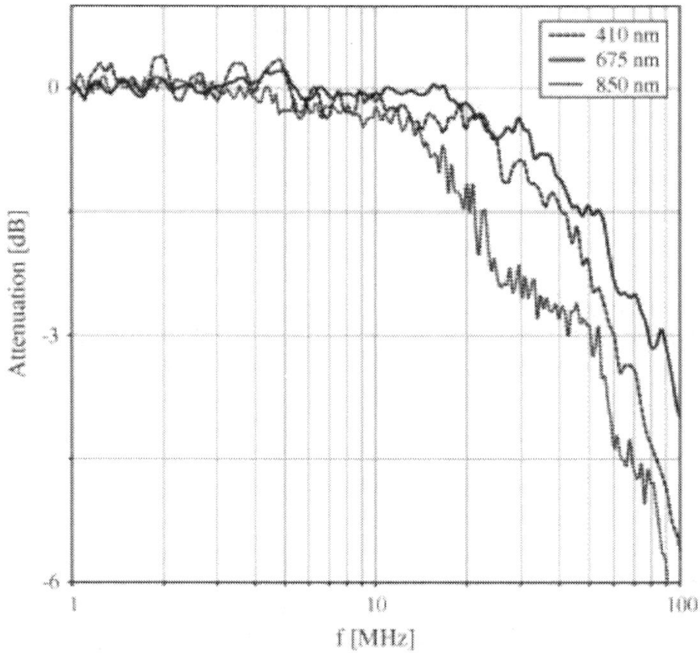

Figure 10. Frequency response of the 40 × 40 μm² $100_B Edge_E$ phototransistor at 410 nm, 675 nm, 850 nm and $V_{CE} = -10$ V.

Table 3. Bandwidths in MHz of the three presented phototransistors at 410 nm, 675 nm, 850 nm as well as floating base and $V_{CE} = -2$ V and -10 V. The top table presents the values for the 40 × 40 μm² phototransistors and the bottom table for the 100 × 100 μm² phototransistors.

	λ = 410 nm		λ = 675 nm		λ = 850 nm	
	$V_{CE} = -2$ V	$V_{CE} = -10$ V	$V_{CE} = -2$ V	$V_{CE} = -10$ V	$V_{CE} = -2$ V	$V_{CE} = -10$ V
$50_B Center_E$	10.7	57.5	9.6	67	12.8	50
$100_B Edge_E$	14.4	60.5	12.1	76.9	14.2	50.7
$100_B Quad_E$	20.2	54.2	18.8	60.3	18.6	31.6
$50_B Center_E$	9.8	36.5	9.1	54	10.5	25.1
$100_B Edge_E$	12.2	40.1	12	58.7	12	25.7
$100_B Quad_E$	16.6	34	16.1	51.6	15.8	21.4

CONCLUSION

In this work we present three types of speed-optimized pnp phototransistors built in a standard 180 nm CMOS process without modifications. Each type of phototransistor was fabricated with areas of $40 \times 40\ \mu m^2$ and $100 \times 100\ \mu m^2$. For achieving high bandwidths a PIN structure was used for the base–collector junction. Hence a special starting material was used consisting of the p^+ substrate and a low doped p^- epi layer grown on top of it. By this low doped epi layer a thick space-charge region is formed, which is necessary for a fast separation of the generated charges caused by deep penetrating light. Since the phototransistors were designed for high-speed applications a further bandwidth increase was achieved by small emitter areas. This emitter area reduction leads also to a reduction of the base-emitter capacitance. However, the small emitter areas are disadvantageous for achieving high responsivities. Thus our phototransistors achieve only small dynamic responsivities up to 2.89 A/W as well as DC responsivities up to 6.44 A/W. Furthermore the phototransistors reach bandwidths up to 50.7 MHz at 850 nm, 76.9 MHz at 675 nm and 60.5 MHz at 410 nm at $V_{CE} = -10$ V and floating base conditions. These results are caused by the small capacitances and the high electric field strengths in the space-charge regions. Furthermore simulations of the electric field strengths and space-charge regions were done. Compared to the phototransistors described in [13] and [14], which were realized in 0.6 μm CMOS technology, the bandwidth is increased by more than a factor of 5. Therefore, these phototransistors are well suited for applications where a high-speed photodetector is needed with an inherent current amplification. A meaningful comparison of the presented phototransistors with other phototransistors is rather difficult since the device is strongly non-linear and its operating conditions are dependent on many factors (e.g. collector–emitter voltage, size of the device, optical light power, wavelength, additional base currents, etc.). However, the authors tried to give a comparison for 850 nm light, shown in Table 4. Possible applications for the presented phototransistors could be for example three dimensional cameras, fast opto-couplers and optical data receivers. Compared to an conventional PIN photodetector using an optimized PIN photodiode (with a responsivity of 0.4 A/W) the presented devices can be used to amplify the input signal up to a factor of 7.2, which equals an optical signal gain in the range of 8.6 dB.

Table 4. Comparison of CMOS and BiCMOS phototransistors at 850 nm.

Refs.	Technology	Device type	Dimension (μm^2)	Wavelength (nm)	P_{opt}(dBm)	Responsivity (A/W)	f_{-3dB} (MHz)	GBW (A/W MHz)
[11]	0.35 μm SiGe HBT BiCMOS	NPN	6 × 10	850	−17	2.7	2000	5400[a]
[12]	0.35 μm SiGe BiCMOS	PNP	21 × 25	850	–	5.2	–	–
[14]	0.6 μm CMOS	PNP	100 × 100	850	−21.2	1.8	14[b]	25.2
[15]	0.35 μm CMOS	PNP	35 × 35	–	–	–	<1	–
[17]	65 nm CMOS	NPN	60 × 60	850	–	0.34	0.15	0.05
This work	0.18 μm CMOS	PNP	100 × 100	850	−15.8	1.5	25.7[b]	38.6
			40 × 40			1.44	50.7[b]	73

[a]Small device illuminated with tapered fiber to get a 2.5 μm diameter optical spot.
[b]Fastest device for 850 nm.

ACKNOWLEDGEMENTS

Funding from the Austrian Science Fund (FWF) in the Project P21373-N22 is acknowledged.

REFERENCES

1. Zimmermann H. Integrated silicon optoelectronics. 2nd ed. Berlin, Heidelberg: Springer-Verlag; 2010.
2. Geist J, Zalewski EF. The quantum yield of silicon in the visible. Appl Phys Lett 1979;35(7):503–6.
3. Christensen O. Quantum efficiency of the internal photoelectric effect in silicon and germanium. J Appl Phys 1976;47:689–95.
4. Schroder DK, Thomas RN, Swartz JC. Free carrier absorption in silicon. IEEE J Solid-State Circuits 1978;13(1):180–7.
5. Swoboda R, Zimmermann H. 11Gb/s monolithically integrated silicon optical receiver for 850 nm wavelength. In: IEEE international solid-state circuit conference, Digest of Technical Papers ISSCC, vol. 49; 2006. p. 240–1.
6. Schaub JD, Li R, Csutak SM, Campbell JC. High-speed monolithic silicon photoreceivers on high resistivity and SOI substrates. IEEE J Lightwave Technol 2001;19(2):272–8.
7. Ciftcioglu B, Zhang L, Zhang J, Marciante JR, Zuegel J, Sobolewski R, et al. Integrated silicon PIN photodiodes using deep N-well in a standard 0.18-lm CMOS technology. IEEE J Lightwave Technol 2009;27(15):3303–13.
8. Davidovic M, Zach G, Schneider-Hornstein K, Zimmermann H. TOF range finding sensor in 90 nm CMOS capable of suppressing 180 klx ambient light. IEEE Sensors 2010:2413–6.
9. Cova S, Ghioni M, Lacaita A, Samori C, Zappa F. Avalanche photodiodes and quenching circuits for single-photon detection. Appl Opt 1996;35(12): 1956–76.
10. Pancheri L, Stoppa D. Low-noise single photon avalanche diodes in 0.15 lm CMOS technology. In: European solid-state device research conference, Proceedings of ESSDERC; 2011. p. 179–82.
11. Yin T, Pappu AM, Apsel AB. Low-cost, high-efficiency, and high-speed SiGe phototransistors in commercial BiCMOS. IEEE Photonics Technol Lett 2006;18(1):55–7.
12. Lai KS, Huang JC, Hsu KYJ. High-responsivity photodetector in standard SiGe BiCMOS technology. IEEE Electron Device Lett 2007;28(9):800–2.
13. Kostov P, Schneider-Hornstein K, Zimmermann H. Phototransistors for CMOS optoelectronic integrated circuits. Sens Actuators, A 2011;172:140–7.
14. Kostov P, Gaberl W, Zimmermann H. Visible and NIR integrated phototransistors in CMOS technology. Solid-State Electron 2011;65–66:211–8.
15. Hu A, Chodavarapu VP. CMOS optoelectronic lock-in amplifier with integrated phototransistor array. IEEE Trans Biomed Circuits Syst 2010;4(5):274–80.

16. Kieschnick K, Zimmermann H, Seegebrecht P. Silicon-based optical receivers in BiCMOS technology for advanced optoelectronic integrated circuits. Mater Sci Semiconduct Process 2000;3:395–8.
17. Carusone AC, Yasotharan H, Kao T. CMOS technology scaling considerations for multi-gbps optical receivers with integrated photodetectors. IEEE J Solid-State Circuits 2011;46(8):1832–42.
18. Kostov P, Gaberl W, Zimmermann H. High-speed PNP PIN phototransistors in a 0.18 lm CMOS process, IEEE ESSDERC 2011. pp. 187–90.
19. Sandage RW, Connelly JA. A fingerprint opto-detector using lateral bipolar phototransistors in a standard CMOS process. IEEE IEDM 1995:171–4.
20. Zhang W, Chan M, Ko PK. A novel high-gain CMOS image sensor using floating N-well/gate tied PMOSFET. IEEE IEDM 1998:1023–5.
21. Winstel G, Weyrich C. Optoelektronik II. Berlin, Heidelberg: Springer; 1986. p. 97.
22. Sze SM, Ng KK. Physics of semiconductor devices. 3rd ed. New York: Wiley; 2006.
23. Kirk CT. A theory of transistor cutoff frequency (fT) falloff at high current densities. IRE Trans Electron Devices 1962:164–74.
24. Whittier RJ, Tremere DA. Current gain and cutoff frequency falloff at high currents. IEEE Trans Electron Devices 1969;16(1):39–57.

CITATION

P. Kostov, W. Gaberl, M. Hofbauer, H. Zimmermann, PNP PIN bipolar phototransistors for high-speed applications built in a 180 nm CMOS process, Solid-State Electronics, Volume 74, August 2012, Pages 49-57, ISSN 0038-1101, http://dx.doi.org/10.1016/j.sse.2012.04.011.

CHAPTER 7

Subband Engineering in *N*-Type Silicon Nanowires Using Strain and Confinement

Zlatan Stanojević, Viktor Sverdlov, Oskar Baumgartner, Hans Kosina

Institute for Microelectronics, TU Wien, Gußhausstraße 27–29, 1040 Wien, Austria

ABSTRACT

We present a model based on $\mathbf{k} \cdot \mathbf{p}$ theory which is able to capture the subband structure effects present in ultra-thin strained silicon nanowires. For electrons, the effective mass and valley minima are calculated for different crystal orientations, thicknesses, and strains. The actual enhancement of the transport properties depends highly on the crystal orientation of the nanowire axis; for certain orientations strain and confinement can play together to give a significant increase of the electron mobility. We also show that the effects of both strain and confinement on mobility are generally more pronounced in nanowires than in thin films. We show that optimal transport properties can be expected to be achieved through a mix of confinement and strain. Our results are in good agreement with recent experimental findings.

HIGHLIGHTS

- We computationally study the electronic structure of strained silicon nanowires.
- We explore the design space for nanowire devices spanned by strain and confinement.
- The device performance can be significantly improved by combining the two effects.
- Our results are in good agreement with recent experimental findings.

INTRODUCTION AND MOTIVATION

Nanowire-based gate-all-around transistors offer a perspective for further device size reduction in microelectronics. On one hand, gate-all-around device architectures exhibit superior electrostatic control of the channel over planar or silicon-on-insulator (SOI) technologies due to a high surface to volume ratio. Improved electrostatic control remedies short channel effects that plague modern planar technologies, especially in the subthreshold regime of operation. On the other hand, if the nanowire channels are made very thin, quantum effects begin to appear. While in traditional planar device architectures quantum effects almost always adversely affect device performance, they offer opportunities for performance improvement in non-planar architectures, such as transistors with nanowire channels.

In a recent experimental study [1], nanowires with gate-all-around structure as thin as 3 nm were successfully fabricated using a top down structuring process [2]. The produced nanowires had a [1 1 0] oriented axis and ($\bar{1}$ 1 0) and (0 0 1) oriented walls. Most notably, the authors presented results of axially strained nanowire field effect transistors where the measured strain-induced current increase surpassed the current increase observed in (1 0 0) oriented thin SOI films in [1 1 0] direction by roughly a factor of two.

In this work we first attempt to examine in detail the subband structure effects causing the current enhancement. Then we explore the design space spanned by nanowire geometry (i.e. quantum confinement) and strain conditions. The outline of this paper is as follows: In Section 2 we will present the $\mathbf{k} \cdot \mathbf{p}$ model of the nanowire subband structure, which accurately treats confinement and strain. Here, we will also give some relevant details about the computational procedure itself. In Section 3 we will present the results of our subband structure calculations. Here, we will discuss the influence of nanowire diameter, cross-section shape, and stress on the subband structure and the transport properties arising therefrom. In Section 4 we will summarize and conclude this paper.

SUBBAND STRUCTURE MODELING

To understand the transport properties in wires less than 10 nm wide, one must carefully take quantization effects into account. A simple treatment using effective masses fails to satisfactorily describe the subband structure of such thin devices. This is due to the energy of the lowest subband already being of the order of 100 meV where non-parabolicity effects become noticeable.

Bulk Hamiltonian

The starting point is the strain-dependent description of the bulk silicon conduction band structure using a two band $\mathbf{k} \cdot \mathbf{p}$ model which is due to Hensel et al. [3]. Commonly in $\mathbf{k} \cdot \mathbf{p}$models, the expansion is taken around the Γ point where the minima and maxima of the interacting conduction and valence bands are located. This is practical for modeling band structures of direct semiconductors and in cases where one looks at the valence band structure only. In silicon, which has an indirect bandgap, this is not feasible. Therefore, the model used here assumes, that the interacting bands are the lowest two conduction bands, with the remaining bands being treated as perturbation. The expansion is performed around one of the three X points where a pair of adjacent Δ valleys touch. The model is valid for the conduction band up to 0.5 eV, as was shown through benchmarking against density functional theory (DFT), semi-empirical tight binding (TB), and empirical pseudopotential method (EPM) calculations [4] and [5]. It includes a first-order treatment of uniaxial and shear strain effects on the conduction band by means of deformation potentials.

The model Hamiltonian describing a pair of adjacent Δ-valleys reads as follows:

$$\mathbf{H} = \begin{pmatrix} \frac{1}{2}\mathbf{p}^{-} \cdot \mathbf{m}^{-1} \cdot \mathbf{p}^{-} + \Xi_{u}\varepsilon_{\zeta\zeta} + V & -\frac{p_{\xi}p_{\eta}}{M} + 2\Xi_{u'}\varepsilon_{\xi\eta} \\ -\frac{p_{\xi}p_{\eta}}{M} + 2\Xi_{u'}\varepsilon_{\xi\eta} & \frac{1}{2}\mathbf{p}^{+} \cdot \mathbf{m}^{-1} \cdot \mathbf{p}^{+} + \Xi_{u}\varepsilon_{\zeta\zeta} + V \end{pmatrix} \quad (1)$$

$$\mathbf{p}^{\pm} = \hbar \begin{pmatrix} k_{\xi} \\ k_{\eta} \\ k_{\zeta} \end{pmatrix} \pm \hbar \begin{pmatrix} 0 \\ 0 \\ k_{0} \end{pmatrix}, \quad \mathbf{m}^{-1} = \begin{pmatrix} m_{t}^{-1} & & \\ & m_{t}^{-1} & \\ & & m_{l}^{-1} \end{pmatrix}.$$

Here, $\mathbf{p}^{\pm}$ are the crystal momenta with respect to the valley minima. $\mathbf{m}^{-1}$ is the inverse of the effective mass tensor at each of the valley minima, with $m_l = 0.91 me$ and $m_t = 0.19 me$. $\frac{1}{M} \approx \frac{1}{m_t} - \frac{1}{m_e}$ describes the coupling between the two valleys; $k_0 = 0.15\frac{2\pi}{a}$ amounts to the distance in k-space between a X point and the Δ valley minima; $\varepsilon\zeta\zeta$ and $\varepsilon\xi\eta$ are uniaxial and shear strain components, and $\Xi_u = 9.0$ eV and $\Xi_{u'} = 7.0$ eV the respective deformation potentials. V denotes the conduction band edge.

The crystal momenta, k-vectors, masses, and strains in (1) are given in the coordinate system of the $\mathbf{k} \cdot \mathbf{p}$ expansion, $\mathbf{e}\xi$, $\mathbf{e}\eta$, $\mathbf{e}\zeta$. One expansion gives a Hamiltonian for only two of the six conduction band valleys in silicon; the Hamiltonian for the other two valley pairs can be obtained by taking even permutations of the basis vectors. The device, which in our case is a nanowire channel, generally uses a coordinate system which differs from the one used in (1); we name it the device coordinate system $\mathbf{e}x$, $\mathbf{e}y$, $\mathbf{e}z$, where $\mathbf{e}z$ denotes the axial coordinate and $\mathbf{e}x$ and $\mathbf{e}y$ are the cross-section coordinates. The transformation matrix reads

$$\mathbf{U}=\begin{pmatrix}\mathbf{e}_\xi\cdot\mathbf{e}_x & \mathbf{e}_\xi\cdot\mathbf{e}_y & \mathbf{e}_\xi\cdot\mathbf{e}_z\\ \mathbf{e}_\eta\cdot\mathbf{e}_x & \mathbf{e}_\eta\cdot\mathbf{e}_y & \mathbf{e}_\eta\cdot\mathbf{e}_z\\ \mathbf{e}_\zeta\cdot\mathbf{e}_x & \mathbf{e}_\zeta\cdot\mathbf{e}_y & \mathbf{e}_\zeta\cdot\mathbf{e}_z\end{pmatrix}. \tag{2}$$

It is convenient to specify the device axes in terms of Miller indices. The main axis of our device is the nanowire axis, which can be specified e.g. as [1 1 0]. If we choose **e***z* as our nanowire axis, the third column in (2) becomes $(1\,1\,0)^T/\sqrt{2}$ in this case. If the wire is not rotationally symmetric the orientation of one of the surfaces must be specified additionally. Using the matrix from (2) we can establish the relations

$$\begin{pmatrix}k_\xi\\ k_\eta\\ k_\zeta\end{pmatrix}=\mathbf{U}\begin{pmatrix}k_x\\ k_y\\ k_z\end{pmatrix},\quad \begin{pmatrix}k_\xi\\ k_\eta\\ k_\zeta\end{pmatrix}^T=\begin{pmatrix}k_x\\ k_y\\ k_z\end{pmatrix}^T\mathbf{U}^T. \tag{3}$$

Using these relations we can express the Hamiltonian (1) in terms of *kx*, *ky*, *kz*, which are defined in the device coordinate system.

Quantization of the Hamiltonian

The next step is to introduce quantization. In a nanowire electrons are only partially quantized: while they are confined within the cross-section of the nanowire, they are free to move along the nanowire axis. This is modeled by substituting the *k*-vector components perpendicular to the axis with derivatives and by parametrizing the axial *k*-vector component.

$$k_x\mapsto -i\frac{\partial}{\partial x},\quad k_y\mapsto -i\frac{\partial}{\partial y},\quad k_z=:k_\parallel;\quad \mathbf{k}_q=\begin{pmatrix}-i\frac{\partial}{\partial x}\\ -i\frac{\partial}{\partial y}\\ k_\parallel\end{pmatrix} \tag{4}$$

The rotated Hamiltonian (1) can be expressed as

$$\mathbf{H}=\begin{pmatrix}\mathbf{k}_q^T\mathbf{D}\mathbf{k}_q & \mathbf{k}_q^T\mathbf{C}\mathbf{k}_q\\ \mathbf{k}_q^T\mathbf{C}\mathbf{k}_q & \mathbf{k}_q^T\mathbf{D}\mathbf{k}_q\end{pmatrix}+\begin{pmatrix}-\mathbf{d}^T\mathbf{k}_q & 0\\ 0 & \mathbf{d}^T\mathbf{k}_q\end{pmatrix}+\begin{pmatrix}H_0^d & H_0^c\\ H_0^c & H_0^d\end{pmatrix}, \tag{5}$$

$$\mathbf{D} = \frac{\hbar^2}{2}\mathbf{U}^T\mathbf{m}^{-1}\mathbf{U}, \quad \mathbf{C} = -\frac{\hbar^2}{2M}\mathbf{U}^T\begin{pmatrix} 0 & 1 & 0 \\ 1 & 0 & 0 \\ 0 & 0 & 0 \end{pmatrix}\mathbf{U}, \mathbf{d} = \frac{\hbar^2}{m_l}\begin{pmatrix} 0 \\ 0 \\ k_0 \end{pmatrix}\mathbf{U}, \quad H_0^d = \frac{\hbar^2 k_0^2}{m_l} + \Xi_u \varepsilon_{\zeta\zeta} + V, \quad H_0^c = 2\Xi_{u'}\varepsilon_{\xi\eta}.$$

At this point the quantization is introduced by adding Dirichlet boundary conditions to the problem, i.e. we require the wavefunction to vanish at the nanowire boundary.

Stress and strain

Strain couples to the Hamiltonian through the deformation potentials Ξ_u and Ξ_u'. In the rotated Hamiltonian the contributions due to strain are summed up in the third term of (5). The relevant strain components $\varepsilon\xi\eta$ and $\varepsilon\zeta\zeta$ are given in the coordinate system of the $\mathbf{k} \cdot \mathbf{p}$expansion, $\mathbf{e}\xi$, $\mathbf{e}\eta$, $\mathbf{e}\zeta$, and are related to the stress tensor σ via the stiffness tensor,

$$\sigma = C\varepsilon, \quad \sigma_{ij} = C_{ijkl}\varepsilon_{kl}, \quad i,j,k,l \in \{\xi,\eta,\zeta\}, \tag{6}$$

which for silicon reads (in engineering notation)

$$C = \begin{pmatrix} C_{11} & C_{12} & C_{12} & & & \\ C_{12} & C_{11} & C_{12} & & & \\ C_{12} & C_{12} & C_{11} & & & \\ & & & C_{44} & & \\ & & & & C_{44} & \\ & & & & & C_{44} \end{pmatrix}, \tag{7}$$

C_{11} = 166.0 GPa, C_{12} = 64.0 GPa, C_{44} = 79.6 GPa [6].

Note, that the strains $\varepsilon\xi\eta$ and $\varepsilon\zeta\zeta$ influence only one of the three valley pairs. Since the Hamiltonians for the other two valley pairs are obtained by rotating the basis vectors, the basis of the strain tensors is rotated as well. Therefore, the relevant strain components for the other two valley pairs are $\varepsilon\eta\zeta$, $\varepsilon\xi\xi$ and $\varepsilon\xi\zeta$, $\varepsilon\eta\eta$, respectively.

In this work we are mainly interested in axial stresses in nanowires. The stress is therefore conveniently given as the stress scalar along the nanowire axis, $\sigma_{\parallel}$. The stress tensor in the crystallographic coordinate system of the $\mathbf{k} \cdot \mathbf{p}$ expansion is constructed as

$$\sigma = \mathbf{U} \begin{pmatrix} 0 & 0 & 0 \\ 0 & 0 & 0 \\ 0 & 0 & \sigma_{\parallel} \end{pmatrix} \mathbf{U}^T. \tag{8}$$

From this the strain tensor is obtained by using (6).

Discretization and numerical solution

The Hamiltonian now has to be discretized in two dimensions (x and y) using an appropriate discretization scheme. In this work we chose box integration because it inherently ensures probability current conservation which is crucial when simulating quantum mechanical systems. It should be noted that the different parts of (5) contribute derivatives of first and second order as well as constant terms. Looking at the symmetric matrices **D**, **C** and the vector **d**

$$\mathbf{D} = \begin{pmatrix} D_{xx} & D_{xy} & D_{xz} \\ D_{yx} & D_{yy} & D_{yz} \\ D_{zx} & D_{zy} & D_{zz} \end{pmatrix}, \quad \mathbf{C} = \begin{pmatrix} C_{xx} & C_{xy} & C_{xz} \\ C_{yx} & C_{yy} & C_{yz} \\ C_{zx} & C_{zy} & C_{zz} \end{pmatrix}, \quad \mathbf{d} = \begin{pmatrix} d_x \\ d_y \\ d_z \end{pmatrix}, \tag{9}$$

we note that the coefficients marked by a solid line are from second order contributions and the coefficients marked by a dashed line are from first order contributions; *Dzz*,*Czz*, and *dz* are due to constant terms.

For each value of the $k_{\parallel}$-parameter, discretizing the Hamiltonian produces a system matrix **A**, the eigenvalues of which are the energies of the subband structure at this particular $k_{\parallel}$ value. To obtain a good approximation of the subband structure the Hamiltonian is discretized for a few hundred values of $k_{\parallel}$. The tasks of diagonalizing the Hamiltonian for different $k_{\parallel}$ values are mutually independent and can therefore be parallelized easily. An example of a calculated subband structure can be seen in Fig. 1 and Fig. 2

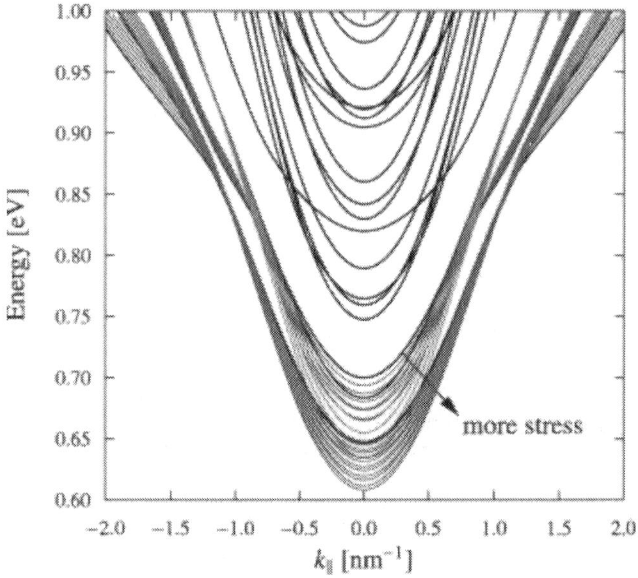

Figure 1. Unprimed subbands in a [1 1 0] nanowire; black lines – unstrained, colored lines – tensile axial stresses up to 1 GPa (shown for the four lowest subbands only). (For interpretation of the references to colour in this figure legend, the reader is referred to the web version of this article.)

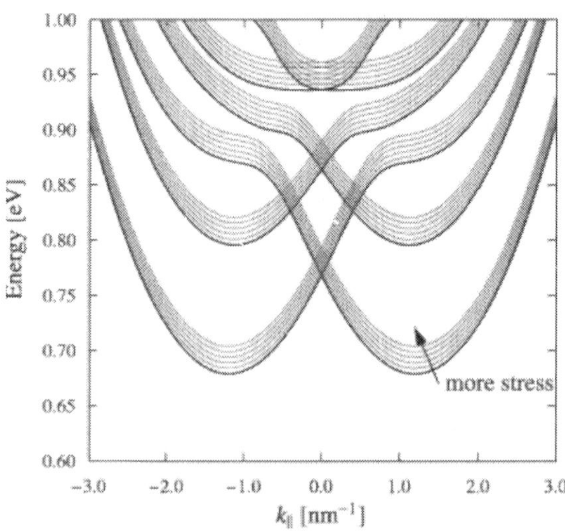

Figure 2. Primed subbands in a [1 1 0] nanowire; black lines – unstrained, red lines – tensile axial stresses up to 1 GPa; plot is centered around the X point at the edge of the Brillouin zone. (For interpretation of the references to colour in this figure legend, the reader is referred to the web version of this article.)

Since we are interested in subbands within the energy range of several k_BT above the valley minimum, computational speed can be further improved if we restrict the calculation of the eigenenergies to the lowest few subbands. The implicitly restarted Arnoldi method (IRAM) provided by the ARPACK library [7] makes use of this restriction and gives sufficiently short computation times to allow the efficient simulation of nanowires up to 7 nm diameter. For larger diameters the energy spacing between the subbands becomes so small that an energy interval of a few k_BT may already contain hundreds of subbands. On the one hand, including more subbands results in a higher computational effort per Arnoldi iteration. On the other hand, closer spaced eigenvalues have a negative impact on convergence speed resulting in more Arnoldi iterations required to find them. This problem can be resolved using a spectral transformation on the matrix called shift-invert [7]. Here a diagonally shifted version of the original system matrix **A** needs to be factorized, i.e. the system

$$(\mathbf{A}-\sigma\mathbf{I})\mathbf{x}=\mathbf{b} \tag{10}$$

needs to be solved for different vectors **b** supplied by the IRAM procedure. The shift-invert transform increases the spacing between the eigenvalues and thus reduces the number required Arnoldi iterations. Also, using multiple shift-invert transformations the total number of subbands to be calculated can be split in several blocks where the number of subbands can be kept low enough for efficient calculation (about twenty subbands per block). The overall computation time then scales linearly with the total number of subbands. This allows us to simulate nanowires with 10 nm diameter and beyond.

Group velocity and effective mass – postprocessing

In the final step we need to extract macroscopically relevant quantities from the subband structure. These are the group velocities and effective masses of the confined electrons, which are related to the subband structure through the first and second derivatives with respect to $k_{\|}$.

In principle it would be possible to calculate the velocities and effective masses by numerical differentiation, i.e. by finite differences. However, this is not particularly advisable because a subband tends to warp significantly especially when energetically close to another subband. A finite difference scheme would require a very fine $k_{\|}$-grid for a reasonably accurate result, especially for the second order derivative.

There is a simple solution to this problem: The group velocity can be calculated using a technique derived from perturbation theory. From there, the effective mass can then be calculated by first order finite differences with satisfactory accuracy. We shall briefly explain the procedure of group velocity calculation in the following.

Non-degenerate perturbation theory states that in presence of a small perturbation $\delta H$the eigenenergy of a state will change according to

$$\delta E_i = \langle\psi_i|\delta H|\psi_i\rangle + \sum_{j\neq i}\frac{|\langle\psi_i|\delta H|\psi_j\rangle|^2}{E_i - E_j} + O(\|\delta H\|^3). \tag{11}$$

We assume δH as the difference in the Hamiltonian between some points $k_{\|}{}^0$ and$k_{\|}{}^0 + \delta k_{\|}$ and approximate it using the $k_{\|}$-derivative of the Hamiltonian to get

$$\delta E_i \approx \left\langle \psi_i \left| \frac{\partial H}{\partial k_{\|}} \right| \psi_i \right\rangle \delta k_{\|}, \tag{12}$$

and obtain the derivative of the eigenenergy,

$$\frac{\partial E}{\partial k_{\|}} = \hbar v_g = \left\langle \psi_i \left| \frac{\partial H}{\partial k_{\|}} \right| \psi_i \right\rangle. \tag{13}$$

One could now proceed to obtain the second derivative by including the second order term in (11). This is impractical, however, because calculating the second order term would require all the matrix elements, or a sufficiently large number of them to keep the error low, whereas the first derivative needs only the diagonal matrix elements, resulting in a much lower computational effort.

In the case of degeneracies in the subband structure the degenerate perturbation theory must be employed. Here we need to distinguish between subbands which are degenerate everywhere, i.e. for all $k_{\|}$, and subbands which are degenerate only on a finite set of $k_{\|}$-points. For the former (13) still holds because moving along $k_{\|}$ does not lift the degeneracy and therefore $\langle \psi i | \partial H / \partial k_{\|} | \psi j \rangle$ vanishes for two degenerate subbands $i$and j. For the latter this is not the case and they need to be treated fully within the degenerate perturbation theory framework.

To obtain the matrix elements we need to calculate the operator $\partial H / \partial k_{\|}$. This is done by analytically differentiating the transformed Hamiltonian (5) with respect to $k_{\|}$ or kz. This gives the matrix

$$\frac{\partial}{\partial k_{\|}}\mathbf{H} = \begin{pmatrix} \frac{\partial}{\partial k_z}\mathbf{k}_q^T\mathbf{D}\mathbf{k}_q & \frac{\partial}{\partial k_z}\mathbf{k}_q^T\mathbf{C}\mathbf{k}_q \\ \frac{\partial}{\partial k_z}\mathbf{k}_q^T\mathbf{C}\mathbf{k}_q & \frac{\partial}{\partial k_z}\mathbf{k}_q^T\mathbf{D}\mathbf{k}_q \end{pmatrix} + \begin{pmatrix} -d_z & 0 \\ 0 & d_z \end{pmatrix}. \tag{14}$$

Using (9) the elements in (14) evaluate to

$$\frac{\partial}{\partial k_{\|}}\mathbf{k}_q^T\mathbf{D}\mathbf{k}_q = 2D_{xz}\left(-i\frac{\partial}{\partial x}\right) + 2D_{yz}\left(-i\frac{\partial}{\partial y}\right) + 2D_{zz}k_{\|}, \tag{15}$$

$$\frac{\partial}{\partial k_{\|}}\mathbf{k}_q^T\mathbf{C}\mathbf{k}_q = 2C_{xz}\left(-i\frac{\partial}{\partial x}\right) + 2C_{yz}\left(-i\frac{\partial}{\partial y}\right) + 2C_{zz}k_{\|}, \tag{16}$$

The matrix elements are calculated by numerically computing the integrals

$$\left\langle \psi_i \left| \frac{\partial H}{\partial k_{\|}} \right| \psi_i \right\rangle = \int \left(\psi_i^{1*} \psi_i^{2*}\right) \frac{\partial}{\partial k_{\|}}\mathbf{H} \begin{pmatrix} \psi_i^1 \\ \psi_i^2 \end{pmatrix} \mathrm{d}A, \tag{17}$$

where the superscript in ψ_i^1 and ψ_i^2 denotes the index of the bulk band. An example of group velocities calculated using this method can be seen in Fig. 3.

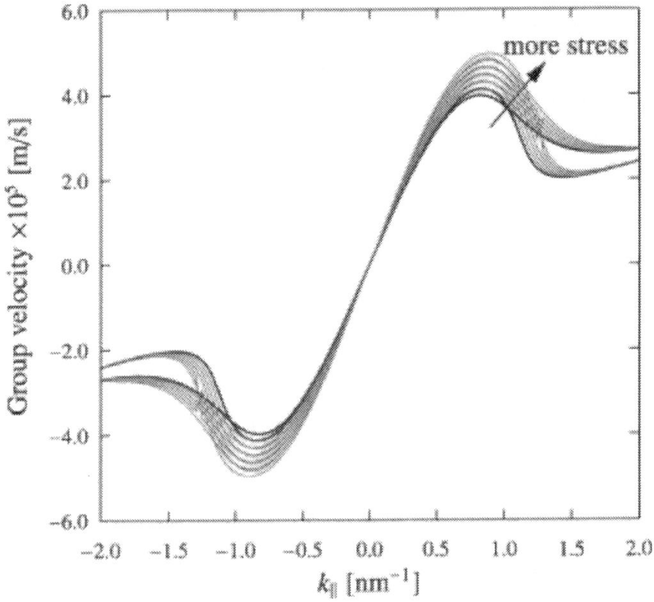

Figure 3. Electron group velocities of the two lowest subbands in a [1 1 0] nanowire (see Fig. 1); black lines – unstrained, red lines – tensile axial stresses up to 1 GPa. (For interpretation of the references to colour in this figure legend, the reader is referred to the web version of this article.)

RESULTS AND DISCUSSION

Using the model and procedure described in the previous section, we simulated nanowires of various diameters assuming square and rectangular cross-sections. The effect of axial stress on the subband structure was mainly investigated in this work. Axial tensile stress is the most likely type of stress to be implemented in a top-down nanowire process, as was demonstrated in [1], and is also a promising technique for mobility enhancement.

Self-consistency was not considered in this work, since we are attempting to isolate the effects of confinement and strain on the subband structure. A positive gate bias would increase the splitting between the unprimed and primed valleys for nanowires thicker than 5 nm. This effect was already studied elsewhere [8] and is beyond the scope of this paper.

Stress behavior of [1 1 0] and [1 1 1] nanowires

Fig. 1 shows the unprimed subbands of a [1 1 0] oriented 5 nm square nanowire. The shape of the subbands is highly non-parabolic which clearly justifies the use of band structure modeling methods beyond the effective mass approximation, such as $\mathbf{k} \cdot \mathbf{p}$. Fig. 2 shows the primed subbands, which are due to valleys that have a lighter quantization mass and therefore lie higher in energy than the unprimed ones. Note, that for [1 1 0] oriented nanowires confinement causes the unprimed subband minima to be folded onto$k_{\|} = 0$, i.e. the Γ point of the one-dimensional Brillouin zone. We shall, therefore, refer to these subbands also as Γ valley.

In both figures the colored lines show the trend of the subband structure change as tensile stress increases up to 1 GPa. The unprimed subbands in Fig. 1 have their valley minima shifted downwards in energy while at the same time their curvature increases. Therefore, we can expect a significant effective mass reduction in the unprimed subbands. In the primed subbands (Fig. 2) tensile stress induces only an upward shift but no change in curvature.

Fig. 4 shows the overall behavior of the confined electrons' effective mass as a function of axial stress for various nanowire thicknesses. We note that both confinement and tensile stress reduces the unprimed subbands' effective mass. Both, however, act in a competitive way: An already low effective mass, due to confinement, undergoes a much smaller change when the nanowire is stressed. For a 12 nm nanowire an axial stress of 1 GPa causes a 22.2% reduction of the effective mass, while in a 3 nm nanowire only 12.2% can be observed. Confinement and [1 1 0] stress both increase the off-diagonal part of the Hamiltonian (1) which explains the competitiveness. As can be seen for all curves in Fig. 4, the impact on the effective mass saturates for high tensile stresses. This saturation effect was already observed in ultra-thin films from calculations using the same model [9]. Apparently the onset of the saturation occurs at lower stresses for smaller diameters due to quantization.

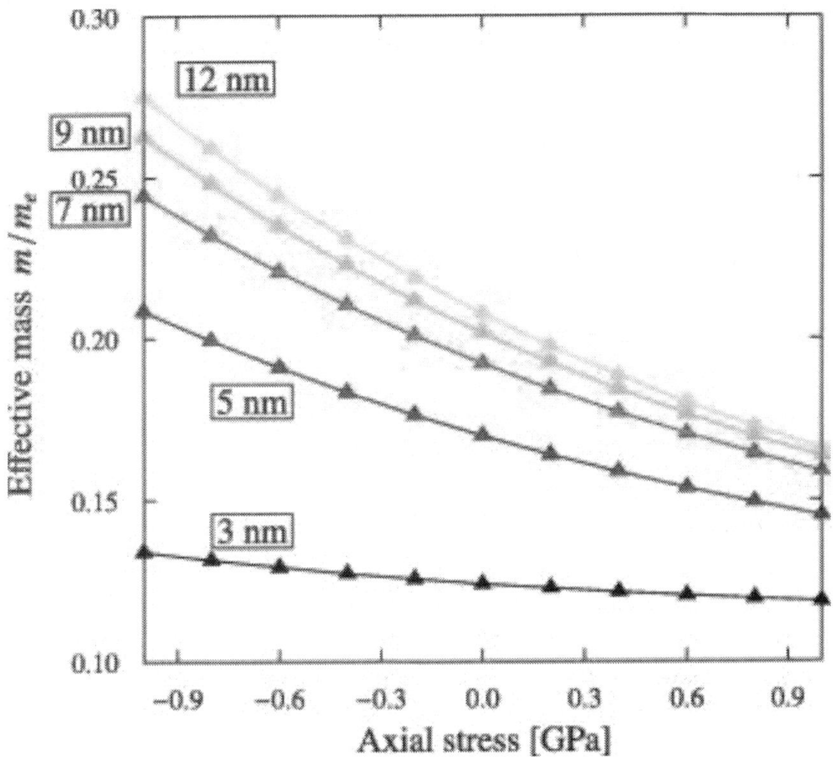

Figure 4. Stress dependence of the effective mass in the unprimed (Γ) valley for [1 1 0] nanowires of different thicknesses.

In Fig. 5 the behavior of the valley minima with respect to axial stress is shown. The unprimed and primed valleys shift in energetically opposite directions when stress is applied, as already mentioned. In this context, tensile stress also benefits the transport properties because it causes a separation of the light unprimed and the heavy primed subbands which effects a higher electron population in the light subbands (see Fig. 6) and a lower intervalley scattering rate due to the energetic remoteness of the primed subbands. We note that for the 3 nm nanowire the unprimed and primed subbands are already far apart due to confinement, so strain will cause no significant improvement, while for the 12 nm nanowire the subband minima almost coincide in energy, which makes application of stress mandatory in order to observe any mobility enhancement.

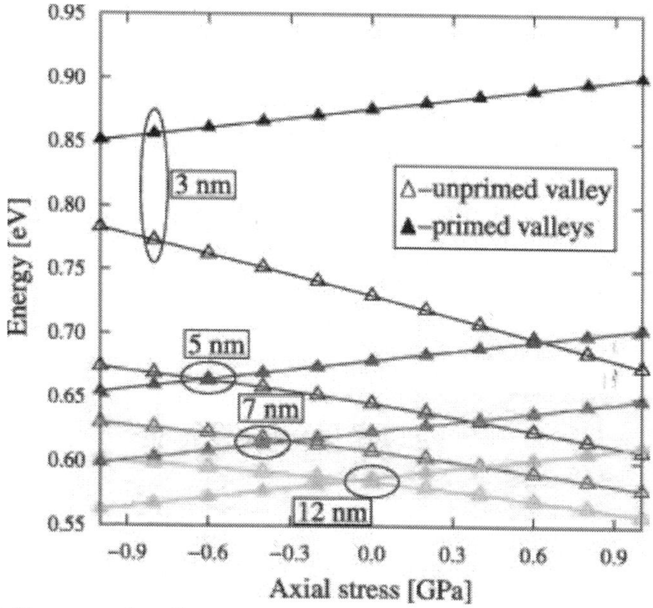

Figure 5. Energy of valley minima for varying axial stress and different thicknesses of [1 1 0] nanowires.

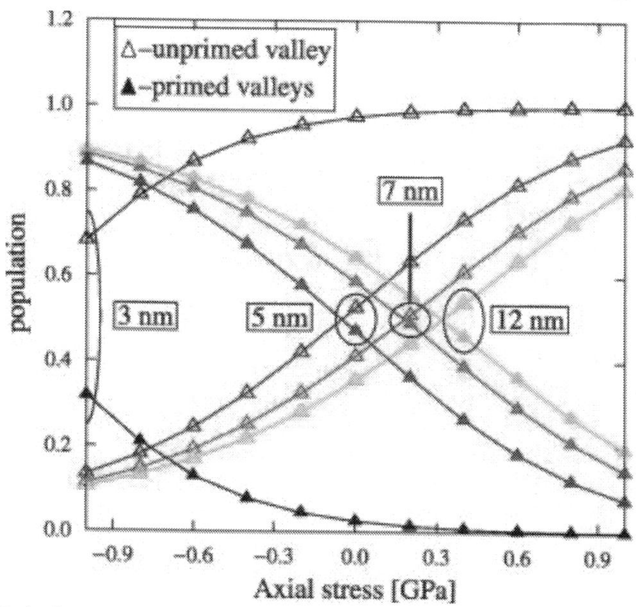

Figure 6. Relative population of unprimed and primed valleys for varying axial stress and different thicknesses of [1 1 0] nanowires.

The situation is entirely different for [1 1 1] nanowires. Here, all valleys have the same quantization mass, and thus no distinction between unprimed and primed subbands is made. As can be seen in Fig. 7, the axial stress merely deforms the subbands without causing any significant shifts or curvature changes in the subband minima. Both confinement and tensile stress cause a splitting of the subbands a the Γ point; the splitting between the first and second subband eventually becomes so large that the two minima merge into a single heavy-effective-mass minimum; this indeed happens for the 3 nm nanowire. The simulation results presented in Fig. 8 and Fig. 9 confirm that axial stress is not beneficial for transport properties in [1 1 1] nanowires.

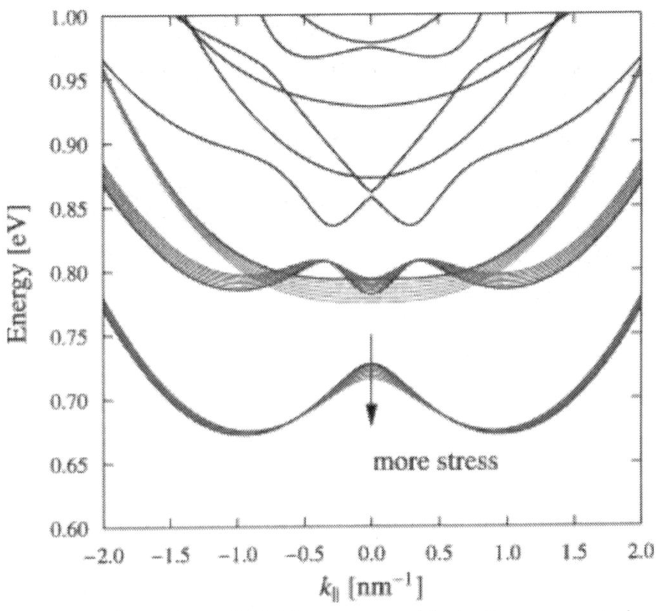

Figure 7. Same as Fig. 1 for a [1 1 1] nanowire; black lines – unstrained, colored lines – tensile axial stresses up to 1 GPa (shown for the three lowest subbands only). (For interpretation of the references to colour in this figure legend, the reader is referred to the web version of this article.)

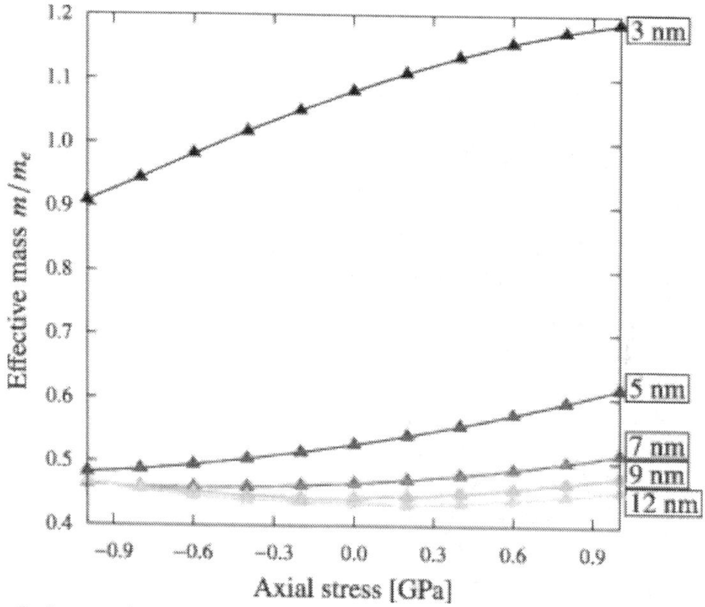

Figure 8. Stress dependence of the effective mass for [1 1 1] nanowires of different thicknesses.

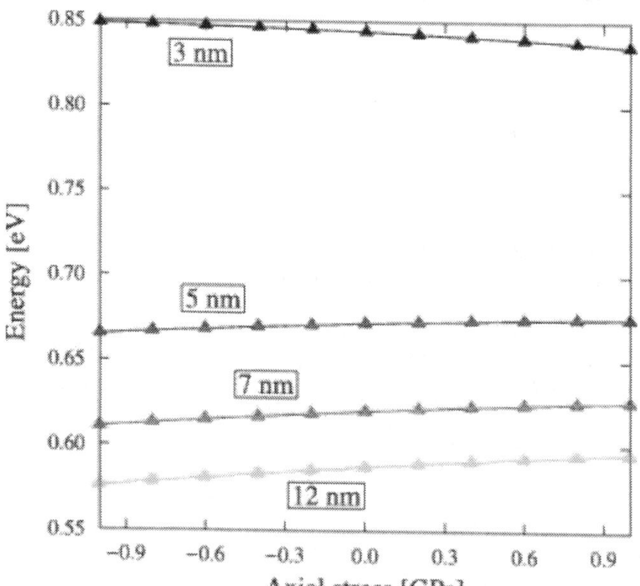

Figure 9. Energy of valley minima for varying axial stress and different thicknesses of [1 1 1] nanowires.

Influence of the aspect ratio on stress behavior

We will now assume the cross-section shape of the nanowires to be rectangular and look at the influence of the rectangle's aspect ratio on the subband structure. We performed several simulations on n-type nanowires of rectangular cross-section while varying the width of the cross-section from 5 nm to 15 nm and keeping the height at 5 nm, similar to recent studies performed for p-type nanowires [10] and [11]. Two sets of simulations were performed assuming the varied surfaces to be $(\bar{1}\,1\,0)$-oriented in one case and (0 0 1)-oriented in the other. Both sets of simulations showed significantly different results, which points out the strong influence of the cross-section shape on the transport properties.

In Fig. 10 and Fig. 11 the stress dependence of the effective mass and valley minima is shown for different widths along the $[\bar{1}\,1\,0]$ axis. The confinement width along $[\bar{1}\,1\,0]$ clearly affects the effective mass and the influence of axial stress upon it. This stems from the fact that by changing the width along any of the $\langle 1\,1\,0 \rangle$ axes one changes the quantization of the off-diagonal coupling elements in (1); these elements are responsible for all the observed effective mass variations. The valley minima also change with the width variation; interestingly, the separation between the unprimed and primed valleys becomes larger as the width increases.

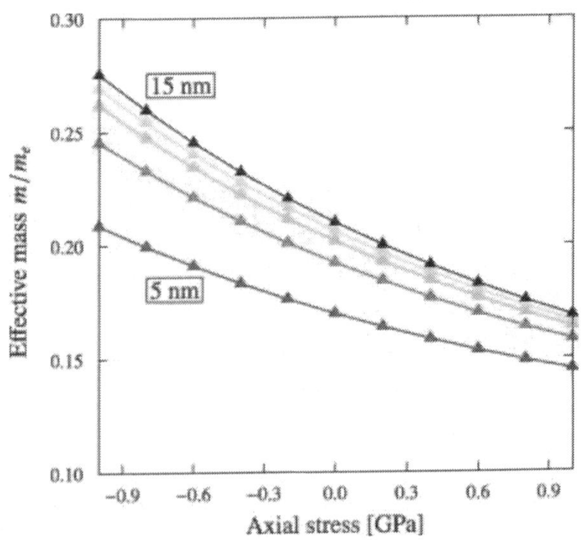

Figure 10. Stress dependence of the effective mass in the unprimed (Γ) valley in [1 1 0] slabs of different widths along $[\bar{1}\,1\,0]$; widths: 5 nm, 7 nm, 9 nm, 11 nm; height: 5 nm; changing the slab thickness along $[\bar{1}\,1\,0]$ clearly affects the effective mass and its dependence on axial strain.

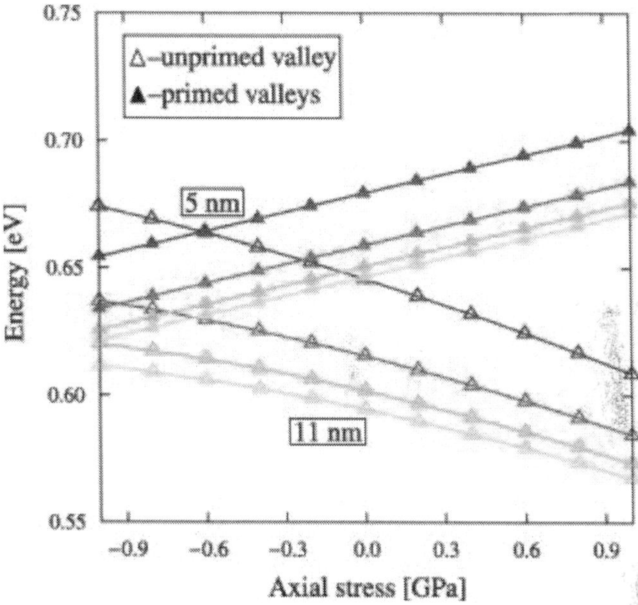

Figure 11. Energy of valley minima of [1 1 0] nanowires for varying axial stress and different widths along $[\bar{1}\,1\,0]$.

In contrast, Fig. 12 shows no change of the effective mass or its stress dependence when the width is changed along [0 0 1]. Since for the Γ valley the width change along [0 0 1] changes the quantization only along the ζ axis in the Hamiltonian (1). The coupling elements and therefore the effective mass remain almost unaffected. The valley minima still respond to the width variations (Fig. 13) but in a different way from what could be seen in Fig. 11; here, the separation between unprimed and primed valleys decreases as the width increases, and for large diameters the primed bands move below the unprimed in the unstrained state.

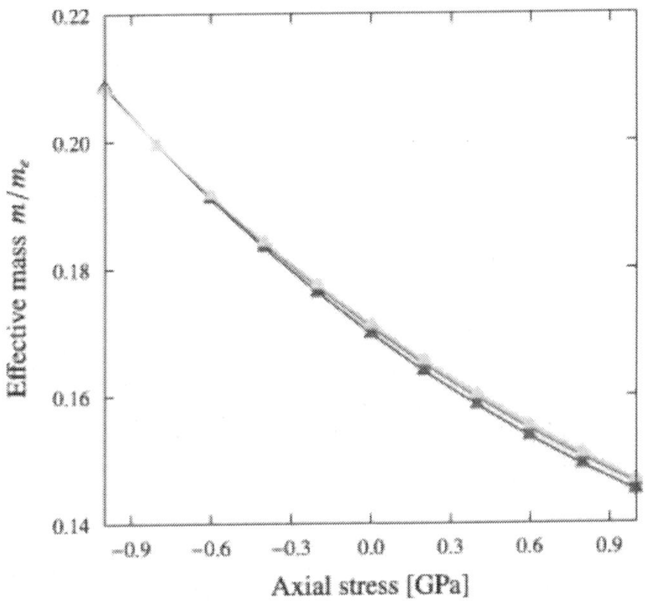

Figure 12. Same as in Fig. 10 but for varying widths along [0 0 1]; here, the curves are on top of each other indicating that confinement along [0 0 1] has no influence on the effective mass in the unprimed valley.

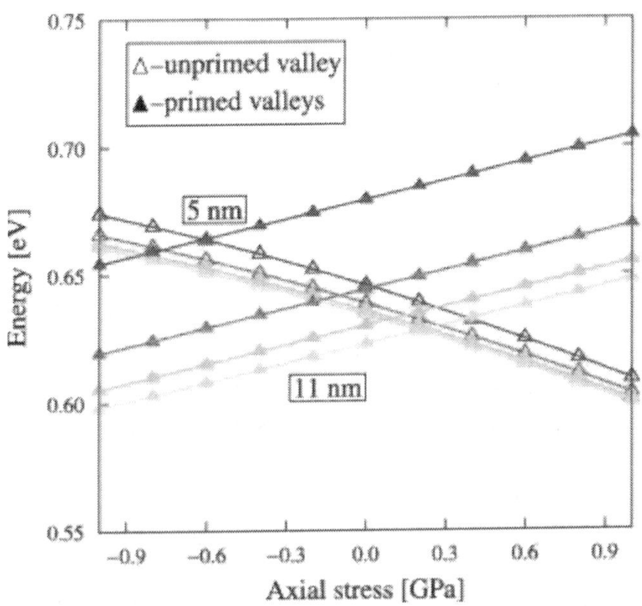

Figure 13. Energy of valley minima of [1 1 0] nanowires for varying axial stress and different widths along [0 0 1].

This lets us conclude that the cross-section shape of [1 1 0] nanowire can be used as an additional design parameter to tune the transport characteristics of nanowires. The dimensions along $[\bar{1}\,1\,0]$ and [001] affect the subband structure in fundamentally different ways. While confinement along $[\bar{1}\,1\,0]$ lets us adjust the effective mass of the confined electrons, the confinement along [0 0 1] allows us to define the energy separation between the unprimed and primed valleys as well as among the subbands within a particular valley, without affecting the transport mass. Therefore, a device optimized for high currents would have a narrow [0 0 1] dimension to keep the subbands well separated and a moderate $[\bar{1}\,1\,0]$ dimension which in combination with tensile axial stress will give a light transport mass.

Comparison of nanowires with thin films

In [1] the authors have found that the on-current increase with tensile stress in 11.2 nm [1 1 0] nanowire *n*-type FETs is about twice as large as for a comparable ultra-thin body (UTB) transistor. This can be attributed to electrons in a nanowire forming a one-dimensional electron gas, whereas in a UTB they form a two-dimensional electron gas. To explain this discrepancy we will employ a simple Drude model to estimate the drift movement of the electrons in nanowire and UTB channels.

$$\langle v \rangle = -\frac{q_0 \langle \tau \rangle}{m_{\mathrm{eff}}} \mathcal{E} \tag{18}$$

Here, $\langle v \rangle$ is the average electron velocity, q_0 is the elementary charge, $\langle \tau \rangle$ is the average momentum relaxation time, m_{eff} is the effective mass, and E is the electric field. The model is, of course, insufficient for an accurate description of the electron movement because it assumes a single parabolic (sub)band. However, the main point here is the dependence of $\langle \tau \rangle$ on the effective mass. In a one-dimensional system the density-of-states mass is the same as the transport mass m_{eff}. If we change the effective mass through confinement or mechanical stress, the electrons both become lighter and scatter less often, since $\tau(E)$ is inversely proportional to the density of statesg$(E\ )$ which for one-dimensional systems is proportional to $\sqrt{m_{\mathrm{eff}}}$, hence $\langle \tau \rangle \propto m_{\mathrm{eff}}^{-1/2}$ and $\langle v \rangle \propto m_{\mathrm{eff}}^{-3/2}$. Indeed, inserting our calculated masses for the 12 nm [110] nanowire gives a velocity enhancement of 27.4% which agrees very well with the 30% found experimentally [1].

In two-dimensional systems the simple relation derived before does not hold because m_{eff} and m_{dos}, the density-of-states mass, are not equal. The density-of-states mass takes the transverse mass into account, which for thin silicon films in [1 1 0] direction is increased by tensile strain, as shown in [9]. The effect of strain on the masses cancels out in the average free flight time and $\langle v \rangle \propto m_{\mathrm{eff}}^{-1}$ resulting in a lower current enhancement due to stress.

SUMMARY AND CONCLUSIONS

In this work we have used a two band **k** · **p** model for the conduction band of bulk silicon and adapted it for *n*-type silicon nanowires. The model provides us with an accurate description of non-parabolicity in the silicon conduction band and includes a treatment of strain effects on the band structure. In the course of the work we have also developed computational methods for dealing with a large number of medium to large-scale eigenvalue problems, i.e. the calculation of the subband structure. The necessary algorithms were implemented within the Vienna Schrödinger Poisson solver framework [12], and allow all the calculations performed in this paper to be executed on a common workstation computer. Furthermore, a method was presented to obtain the exact value of the group velocities at a certain point in *k*-space without numerical differentiation of the subband structure. Skipping one numerical differentiation step improves the accuracy of the effective mass extraction from the subband structure. Its usability is not restricted to our calculations but can be used in other **k** · **p** models and can in principle be extended to any (sub)band structure calculation method relying on the repeated diagonalization of a *k*-dependent Hamiltonian, such as tight binding or pseudopotential methods.

With the presented simulation framework we have studied the behavior of axially stressed silicon nanowires of various thicknesses. We have shown how confinement and stress act on the electron subband structure of nanowires and pointed out where performance improvement can be expected. The effect of different aspect ratios of the nanowire cross-section on the subband structure properties was investigated for [1 1 0] nanowires. It was found that confinement along the $[\bar{1}\,1\,0]$ and [0 0 1] axes has profoundly different effects on the subband structure. It was made clear that diameter, cross-section shape, and stress offer an additional design space for future devices based on [1 1 0] nanowire channels. This allows tuning the device for a particular application through geometrical patterning and application of stress during the fabrication process.

Finally, we addressed the question why the measured current enhancement of stressed nanowires is significantly larger than in thin films under the same stress conditions [1]. According to our results, the effect can be attributed to the one-dimensional nature of the electrons confined in a nanowire. Using a simple transport model for the confined electrons we calculated a stress-induced current enhancement figure close to the one obtained experimentally.

ACKNOWLEDGMENT

This work has been supported by the Austrian Science Fund, special research program IR-ON (F2509).

REFERENCES

1. Bangsaruntip S, Majumdar A, Cohen G, Engelmann S, Zhang Y, Guillorn M, et al. Gate-all-around silicon nanowire 25-stage CMOS ring oscillators with diameter down to 3 nm. In: Symposium on VLSI technology, 2010 (VLSIT, 2010); 2010. p. 21–2. doi:10.1109/VLSIT.2010.5556136.

2. Bangsaruntip S, Cohen G, Majumdar A, Zhang Y, Engelmann S, Fuller N, et al. High performance and highly uniform gate-all-around silicon nanowire MOSFETs with wire size dependent scaling. In: IEEE international electron devices meeting (IEDM, 2009); 2009. p. 1–4. doi:10.1109/IEDM.2009.5424364.
3. Hensel JC, Hasegawa H, Nakayama M. Cyclotron resonance in uniaxially stressed silicon. II: nature of the covalent bond. Phys Rev 1965;138(1A):A225–38. doi:10.1103/PhysRev.138.A225.
4. Sverdlov V, Karlowatz G, Dhar S, Kosina H, Selberherr S. Two-band k p model for the conduction band in silicon: impact of strain and confinement on band structure and mobility. Solid-State Electron 2008;52(10):1563–8.
5. Sverdlov VA, Windbacher T, Schanovsky F, Selberherr S. Mobility modeling in advanced MOSFETs with ultra-thin silicon body under stress. J Integr Circ Syst 2009;4(2):55–60.
6. Levinstein M, Rumyantsev S, Shur M. Handbook series on semiconductor parameters. London: World Scientific; 1996. vol. 1. 2, 1999, 191.
7. Lehoucq R, Sorensen D, Yang C. ARPACK users' guide: solution of large-scale eigenvalue problems with implicitly restarted arnoldi methods; 1998.
8. Neophytou N, Paul A, Lundstrom MS, Klimeck G. Self-consistent simulations of nanowire transistors using atomistic basis sets. In: Grasser T, Selberherr S, editors. Simulation of semiconductor processes and devices 2007. Vienna: Springer; 2007. p. 217–20. doi:10.1007/978-3-211-72861-1_51, URL http://dx.doi.org/10.1007/978-3-211-72861-1_51.
9. Sverdlov V, Baumgartner O, Windbacher T, Schanovsky F, Selberherr S. Thickness dependence of the effective masses in a strained thin silicon film. In: International conference on Simulation of semiconductor processes and devices, 2009 (SISPAD '09); 2009. p. 1–4. doi:10.1109/SISPAD.2009.5290252.
10. Neophytou N, Kosina H. Large enhancement in hole velocity and mobility in ptype
1. 110. and
2. 111. silicon nanowires by cross section scaling: an atomistic analysis. Nano Lett 2010;10(12):4913–9. doi:10.1021/nl102875k, http://dx.doi.org/10.1021/nl102875k.
11. Neophytou N, Klimeck G, Kosina H. Subband engineering for p-type silicon ultra-thin layers for increased carrier velocities: an atomistic analysis. J Appl Phys 2011;109(5):053721. doi:10.1063/1.3556435.
12. Karner M, Gehring A, Holzer S, Pourfath M, Wagner M, Goes W, et al. A multipurpose Schrödinger–Poisson solver for TCAD applications. J Comput Electron 2007;6(1):179–82.

CITATION

Zlatan Stanojević, Viktor Sverdlov, Oskar Baumgartner, Hans Kosina, Subband engineering in n-type silicon nanowires using strain and confinement, Solid-State Electronics, Volume 70, April 2012, Pages 73-80, ISSN 0038-1101, http://dx.doi.org/10.1016/j.sse.2011.11.022.

CHAPTER 8

The Different Roles of Charged and Neutral Atomic and Molecular Oxidising Species in Silicon Oxidation from AB Initio Calculations

M.A Szymanski[1], [2], A.M Stoneham[1], A Shluger[1]

[1] Department of Physics and Astronomy, University College London, Gower Street, London WC1E 6BT, UK

[2] Faculty of Physics, Warsaw University of Technology, ul. Koszykowa 75, 00-662 Warsaw, Poland

ABSTRACT

We examine the roles of charged and neutral oxidising species based on extensive ab initio DFT calculations. Six species are considered: interstitial atomic O, O^-, O^{2-} and molecular species: O_2, O_2^-, O_2^{2-}. We calculate their incorporation energies into bulk silicon dioxide, vertical electron affinities and diffusion barriers. In our calculations, we assume that the electrons responsible for the change of charge state come from the silicon conduction band, however the generalisation to any other source of electrons is possible and hence our results are also relevant to electron-beam assisted and plasma oxidation. The calculations yield information about the relative stability of oxidising species, and the possible transformations between them and their charging patterns. We discuss the ability to exchange O atoms between the mobile species and the host lattice during diffusion, since this determines whether or not isotope exchange is expected. Our results show very clear trends: (1) *molecular species* are energetically preferable over*atomic* ones, (2) *charged species* are energetically more favourable than *neutral* ones, (3) diffusion of *atomic species* (O, O^-, O^{2-}) will result in oxygen exchange, whereas the diffusion of *molecular species* (O_2, O_2^-, O_2^{2-}) is not likely to lead to significant exchange with the lattice.

Our results show thermodynamic trends for oxidising species to capture electrons from Si during oxidation. We identify very different roles for atomic and molecular species and also for different charge states of those species. This points out to opportunities, usually not considered, for optimising thin oxide layers and interface properties for use in electronics devices.

INTRODUCTION

From the beginning of silicon technology there has been pressure to scale down the size of devices. This has led to requirement for growing high quality silicon dioxide with a very low concentration of intrinsic defects, which implies a need for better control and understanding of molecular processes during silicon oxidation. This work addresses question of character and roles of oxidising species and charging processes relevant to ultra-thin oxide layers.

Modern gate-dielectrics approach a thickness of a few nm, for which the deviations from classical linear-parabolic (Deal–Grove) theory [1] of oxidation are significant. Thermal oxidation at lower temperatures and lower pressures [2], electron-beam assisted oxidation [3], as well as plasma oxidation [4] cannot be described within linear-parabolic regime. One idea has been that charged oxidising species lead to different kinetics due to the electrostatic field within the oxide layer and image forces arising from difference in the dielectric constant of silicon and silicon dioxide. The effects of image forces were examined by Stoneham and Tasker [5] and the electric field within oxide and ionic conductivity in this field were the basis for power-parabolic ionic oxidation model [6].

The role of charged species in silicon oxidation has been stressed previously several times. Early stopping field experiments [7] have shown that oxidation can be enhanced, retarded or even stopped by applying bias voltage across the sample. More recent experiments [8] and [9] suggested ionic contribution to the transport processes in SiO_2. The kinetics of the oxidation process has been shown to be strongly affected by the presence of a low energy electron-beam, and an anomalous temperature dependence has been observed for oxides less than 2 nm thick [3]. The initial oxidation at low temperatures has been shown to be enhanced by electron impact with the possible dissociation of adsorbed O_2 molecules to form O and O^- species [10]. Plasma oxidation kinetics was observed to follow trends expected for ionic species [4]. Cw-UV induced oxidation was shown to be consistent with model based on photoinjection of electrons from Si resulting in ionic species [11]. Oxygen absorption on SiO_2 surface has been reported to result in O_2^- species [12]. Also other phenomena like telegraph noise [13] and fixed oxide charge point to charge transfer processes as possible causes.

ATOMIC-SCALE PROCESSES IN SILICON OXIDATION AND BASIC MODELS

It is generally believed that diffusion of oxidising species and interfacial processes are the main factors controlling oxidation rate. Fig. 1shows the main oxidation processes on the example of oxygen molecule being the oxidising species. The species sticks to the surface from the gas phase, incorporates into the silica lattice, diffuses toward Si/SiO_2interface and reacts with silicon. This is the underlying picture of the most classical model of silicon oxidation – the Deal–Grove model. This model, however, cannot account for proper kinetics of silicon oxidation for low temperature and pressures or for thin oxides [2]. Plasma or electron-beam assisted oxidation are totally beyond the scope of Deal–Grove model. This simple picture of oxidation can be complicated by adding other intermediate stages.

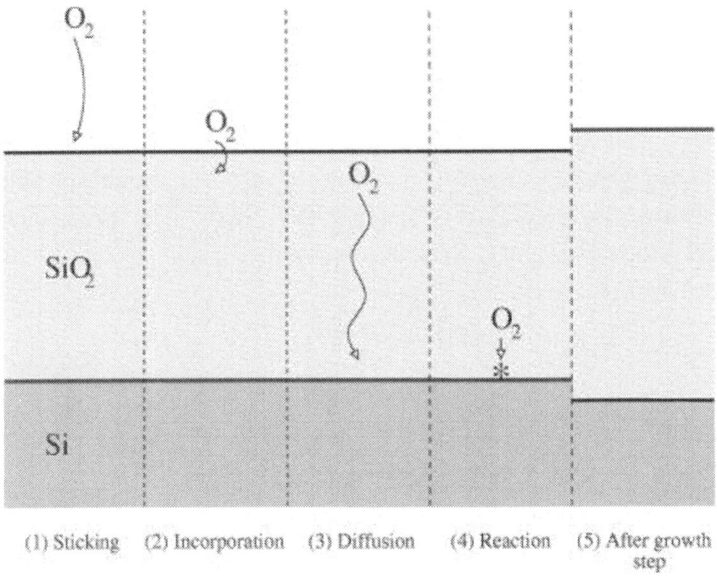

Figure 1. The main oxidation steps on a example of neutral oxygen molecule being the oxidising species. Adsorption from the gas phase (1), followed by incorporation into bulk SiO_2 (2), diffusion towards Si (3), and the reaction at the Si/SiO_2 interface (4), resulting in expansion of the volume of silica (5).

- The oxidising species could react with the silica network exchanging oxygen atoms. This however is in disagreement with isotope exchange experiments [14] which show that oxygen exchange takes place only close to the two interfaces.
- Defects could be generated to accommodate the lattice mismatch at the Si/SiO_2interface (the volume per Si atom in silica is twice that in c-Si).
- There can be many different kinds of charge transfer processes involved. For example, electrons can be transferred from Si to oxidising species which are either adsorbed at the ambient/silica interface or already incorporated in the bulk SiO_2. Those two processes are shown schematically at Fig. 2 and Fig. 3. The former is more likely to happen for thin oxide layers for which a direct tunnelling from Si is possible, the latter could happen within tunnelling range from the Si/SiO_2 interface.

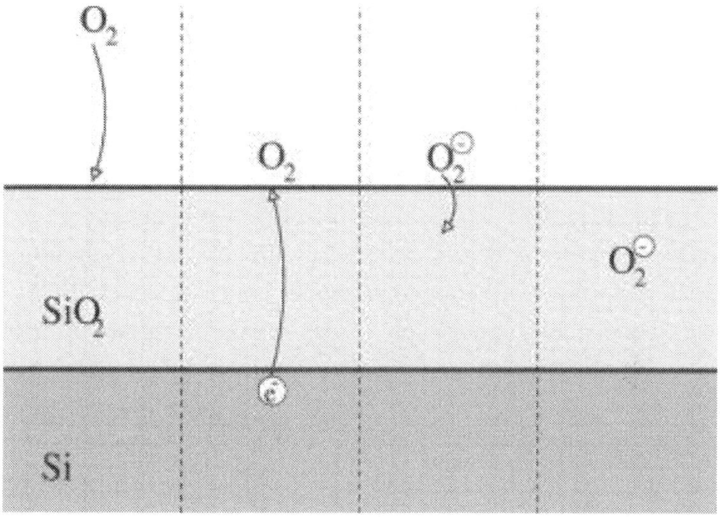

Figure 2. Electron transfer to species adsorbed at the ambient/silica interface. This process is likely for initial stages of oxidation for which direct tunnelling can be efficient.

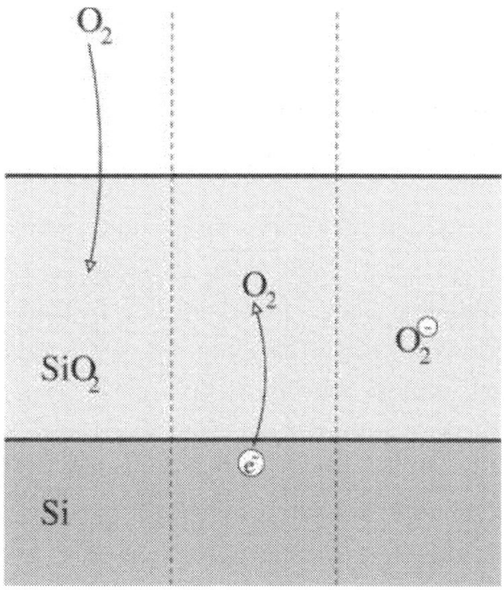

Figure 3. Electron transfer from Si to oxidising species already incorporated into bulk SiO_2. This process can be important for species within electron tunnelling range from Si.

TECHNIQUES

Our calculations use spin-polarised version of density functional theory (DFT) with the Perdew–Wang functional and the generalised gradient approximation (GGA)[15] and [16] implemented within VASP code [17] and [18]. The Kohn–Sham orbitals were expanded in plane-wave basis set with a cutoff energy of 400 eV. Oxygen atoms were represented by `ultra-soft' pseudopotentials, and silicon atoms by norm-conserving ones. Our reference structure is 72-atom fully relaxed hexagonal periodic unit cell of α-quartz. For other calculations, the lattice vectors were fixed and all atoms were allowed to relax. Defects are separated by more than 10 Å. The Brillouin zone integration includes the Γ-point only. We use a neutralising background for charged unit cells, with the Makov–Payne monopole–monopole energy correction [19]. This procedure and our large unit cell minimised the influence of the periodic images of the defects, and thus ensures that our results provide good models of isolated defects in a non-defective host lattice.

The geometry relaxation used conjugate gradient (CG) energy minimisation, with two search methods for the transition state. The ascending and descending valley points method combined with CG energy minimisation gave two estimates of the transition state position on both sides of the saddle point. The join of those two points was then assumed to be a good approximation to the direction of the transition state and the system geometry was driven towards the saddle point by reversing the force along this direction and using a semi-Newtonian relaxation algorithm. The transition states found are configurations of maximum energy along the lowest energy path joining two energy minima on the many dimensional energy surface. All relaxation procedures were terminated when forces on atoms were smaller than 0.05 eV/Å.

The analysis of charged species energies in a periodic model is complicated. A neutralising background is needed to converge Ewald summations. This introduces errors into the total energy expression which can be handled in first order by the Makov–Payne monopole–monopole energy correction, which improves the convergence of the total energy with respect to the size of the unit cell. In our calculation, when an electron was added or removed from the system, the electron came from or went to the zero energy level on the eigenvalue scale. This zero energy level is not normally the experimental vacuum level for bulk calculations, since there is no surface which would determine the position of band levels with respect to the vacuum level. Also, the Ewald sum is defined up to a constant which can affect the position of zero energy. However, in our VASP calculation zero energy is a good reference level since it stays in the same position with respect to the SiO_2 valence band regardless of the type of species and its charge state. This in turn means that all results with energy converged with respect to the size of the unit cell correspond to an isolated point defect in an infinite non-defective crystal. The charging electrons are assumed to come from the bottom of silicon conduction band at the Si/SiO_2 interface. We can use the experimental information on the band offset at the interface [20] and [21] (4.6 eV for Si/SiO_2 valence band offset, 1.1 eV for band gap of Si) and the position of the top of valence band of SiO_2 in our calculations to estimate the energy of an electron E_{el} at the bottom of the conduction band of Si with respect to the theoretical zero energy level [22]; it is easy to adjust energies for another source of electrons.

The only meaningful comparisons are of energies between systems with the same number of electrons. The incorporation energies are calculated for lowest energy configuration of the system with respect to the non-defective quartz structure and an isolated oxygen molecule:

$$E_{inc} = E_{system} - (E_{quartz} + fE_{O_2} + nE_{el})$$

where f=1 for molecular oxygen incorporation and f=0.5 for incorporation of atomic species; n is the number of excess electrons. The vertical electron affinities are calculated by adding an electron to the lowest energy configuration of the system and looking at the energy change without allowing for any further relaxation:

$$E_{aff} = -E_{charged,unrelaxed} + (E_{system} + E_{el})$$

We should note that our model is approximate. In particular, we neglect the band bending, image interactions and electric field build-up due to redistribution of charge associated with charging of oxidising species. However, we use the experimental information about the band offset at the interface which should account partly for those effects. The supercell calculation does not include all of the long-range polarisation effects which means that full treatment of polarisation would favour the charged species even more.

RESULTS FOR OXIDISING SPECIES IN QUARTZ

We have calculated properties of six oxidising species in alpha-quartz which we regarded in first approximation as an acceptable mimic of amorphous silica. The typical relaxed geometries of the species are shown in Fig. 4 and Fig. 5. Our results are summarised in Table 1, in which a positive value of incorporation energy means that this energy has to be supplied to the system in order to incorporate the species into the lattice. The diffusion barrier gives the lowest possible energy barrier which a species encounters during diffusion from one energy minimum to another. We quote the diffusion barriers calculated along the *c* axis of alpha-quartz; diffusion barriers along other directions were calculated only for some species. A positive value of the vertical electron affinity means that the species can capture an electron from the bottom of Si conduction band by resonant tunnelling. This value is calculated for the lowest energy configuration of the species. Some important conclusions can be drawn on the basis of our results:

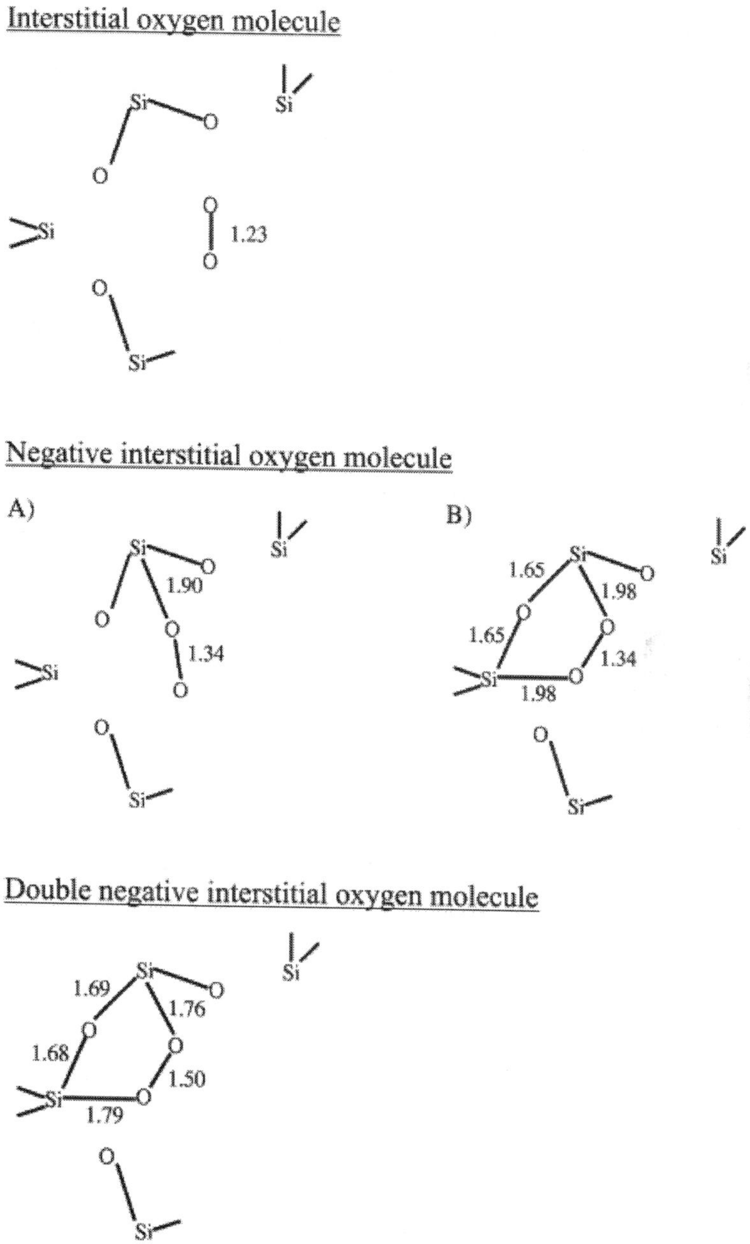

Figure 4. Structures of molecular oxygen species in quartz.

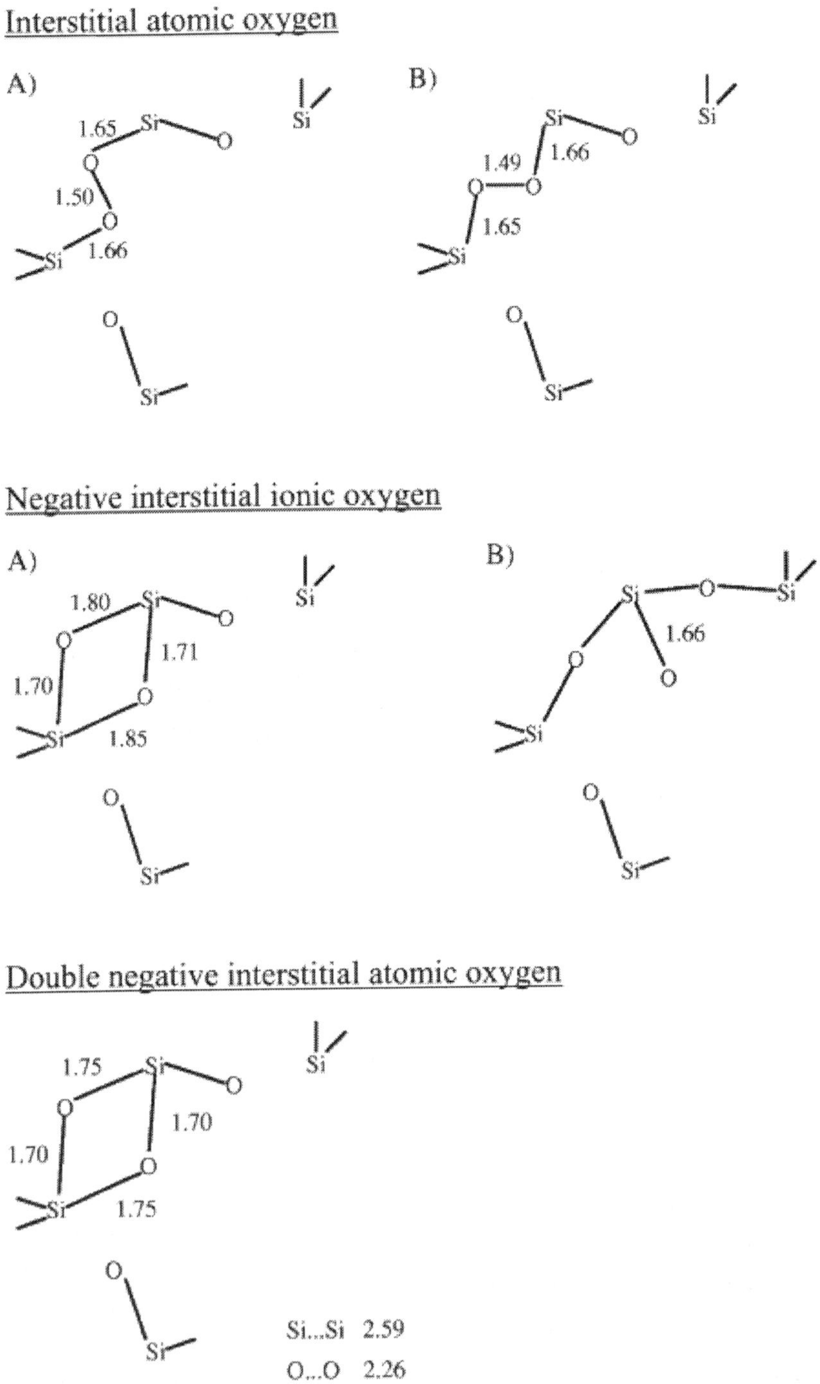

Figure 5. Structures of atomic oxygen species in quartz.

Table 1. Key energies for different species in α-quartz. Extra electrons are assumed to be taken from the bottom of the Si conduction band[a].

Species	Incorporation energy (eV)	Diffusion barrier along *c* (eV)	Diffusion barrier other axes (eV)	Electron affinity (eV)	Exchange with lattice
O_2	2.07	0.09		0.7	No
O_2^-	−0.5	0.57		−0.2	No
O_2^{2-}	−1.9	2.1		No affinity	No
O	2.03	1.3	1.42	−0.5	Yes
O^-	−0.7	0.19	0.6	−0.3	Yes
O^{2-}	−2.8	0.19		No affinity	Yes

[a]Positive incorporation energy means that energy must be supplied to incorporate the species from the gas phase. A positive vertical (constant atomic positions) electron affinity means exothermic capture of an electron from the bottom of the bulk Si conduction band. For all quoted species the charge and spin densities are localised on the species with the exception of neutral O and doubly bounded O^- for which the interstitial and lattice oxygens are in the same charge state (which implies some charge transfer from the lattice to the interstitial oxygen). Therefore, those two species should be thought as complexes. The third excess electron cannot be localised on the species and hence the double negative species do not exhibit any electron affinity.

- *Molecular species* are energetically more favourable than *atomic species*, in line with earlier views on oxidation. For example, the incorporation energy of two interstitial atomic oxygens is 4.1 eV whereas incorporation energy of oxygen molecule is 2.1 eV, which means that dissociation of interstitial oxygen molecule into two atomic species requires 2.0 eV of energy and that atomic oxygens react exothermically to form molecules. However, interstitial atomic oxygen once incorporated into lattice may never encounter another oxygen to react with (given the low oxygen solubility in silica). As a consequence, for instance, it can be expected that plasma oxidation happens mainly due to diffusion of atomic species (possibly in negative charge state). As long as the comparison of atomic and molecular species is done for systems with the same amount of excess charge and number of oxygens the conclusions are general and do not depend on the position of the electron source in our system.

- *Charged species* are energetically more favourable than the *neutral species* (with assumption that electrons are available at the bottom of the conduction band or above).The incorporation energy decreases with the increasing negative charge state (compare the incorporation energies in Table 1). Polarisation of silica is only partially included in supercell calculations and the part missing will favour charged species even more. Charging should happen if there is an efficient way of transferring an electron from Si to the species incorporated in silica, for instance if the species stays within an electron tunnelling distance from Si for sufficient time. The electron affinities give the energy gain after species in its lowest energy configuration captures an electron form the bottom of Si conduction band without relaxation. During diffusion the species will sample a number of other higher energy configurations for which the electron capture may be more or less likely and therefore electron affinities quoted in Table 1 should be regarded only as rough estimates. Although it is energetically preferable to have negative charge localised on oxidising species rather than in form of conduction electrons in Si, in some cases it may require inelastic tunnelling [23] to transfer the electron, as indicated by negative vertical electron affinities.
- The mechanism of diffusion of atomic species will lead inevitably to oxygen exchange with the lattice whereas diffusion of molecular species cannot easily lead to oxygen exchange with non-defective SiO_2 lattice (as discussed in Ref. [24]). It is likely that the high barrier for motion of O_2^{2-} will prevent the diffusion of its molecular form and the dissociation into two O^- seems to be a strong possibility (although we did not calculate the energy barrier for this process).
- Capturing an electron the molecular species can either relax in molecular form or dissociate into two atomic species in appropriate charge states. The former process is always more likely from thermal equilibrium point of view. It does not eliminate, however, the possibility of dissociation. Isotope exchange experiments during dry oxidation [14] show significant exchange only close to both interfaces. This exchange can be due to creation and motion of atomic species; in

particular, near the Si/SiO_2 interface due to electron transfer form Si to oxidising species with some of them dissociating into atomic ones. If so, there may be no need for so called "reactive layer" near the Si/oxide boundary. The different properties of this layer could be explained by charging which changes the character of the oxidising species and dissociation of some of them after electron capture.

OXIDISING SPECIES IN AMORPHOUS SiO_2

Mott [25] pointed out that oxygen solubility data for silica can be only understood if there is small number favoured energy sites. This is obviously not so in quartz. The experimentally observed activation energy for oxidation (oxygen incorporation energy plus diffusion energy) is 1.2 eV. The calculated activation energy for oxidation by neutral O_2 in quartz is 2.2 eV. This suggests that crystalline SiO_2 could be a poor mimic of silica for some of the species. Our preliminary results for amorphous structures [26] show that we can expect significantly different results only in case of O_2 and O_2^-; incorporation energies and electron affinities for other species are generally similar. For O_2 the average incorporation energy at three different interstitial sites in an amorphous sample is 0.4 eV (2.1 eV for quartz), for O_2^--1.2 eV (−0.5 eV for quartz). There is also a relatively small number of low energy sites for the neutral interstitial oxygen molecule, in agreement with Mott`s prediction. As it can be seen from Fig. 4 the interstitial site for O_2 is in the middle of a void. Our calculations in amorphous sample show that the bigger the void in which oxygen molecule is placed, the lower the incorporation energy. The amorphous structure offers a number of much bigger voids than quartz and hence the incorporation energy is much lower. The diffusion barrier on the other hand will depend on the size of ring through which the molecule travels. The smaller the ring the bigger the barrier. From this perspective, diffusion of O_2 along the *c* channel of quartz is like a travel through a tunnel of fixed diameter and hence the very low diffusion barrier. The diffusion in amorphous silica will involve hops between relatively big voids through narrower rings. Therefore we can expect diffusion barriers to be higher than in the case of quartz. If diffusion barriers are of order 0.8 eV, we get an excellent agreement with experimental activation energy for oxidation. Clearly, appropriate description of molecular species in amorphous silica requires some level of static disorder. However, despite the differences for the molecular species in quartz and amorphous structures all conclusions from previous section hold also for oxygen species of amorphous silica.

ROLES OF ATOMIC AND MOLECULAR SPECIES IN OXIDATION AND THE CONSEQUENCES OF CHARGING

The two types of species: atomic and molecular may have very different roles in the oxidation processes (their incorporation energies and diffusion patterns are different). They may react with different sites at the interface and lattice defects. This may affect both the final oxide structure as well as the number of interfacial and bulk defects and, as a result, the breakdown reliability. Experiments show different properties (density and distribution of Si atoms) of the near-interfacial region for oxidation in dry O_2 and ozone [27]. It can be expected that ozone provides atomic oxygens during oxidation and hence changes ratio of atomic to molecular species arriving to the interface, which supports our predictions.

Charging of the oxidising species may have important consequences for the structure of the oxide and the Si/SiO_2 interface. The fact that incorporation energies and diffusion mechanisms differ for different charge states is likely to affect the reactivity of the species and change the potential reaction sites. Electron transfer will build up an electric field in silica and this field, in turn, will influence the spatial distribution and diffusion of charged species. A localised charge will also induce polarisation of Si which will result in an attractive force on the oxidising species towards Si. Applying a bias voltage across the sample can promote or prevent drift of the charged species and their redistribution resulting in different ratio of different charge species arriving to the Si/SiO_2 interface. The role of charged species identified here is different from suggested by other authors.

What we have shown is that different oxidising species have a variety of behaviours which can be exploit during silicon oxidation for optimisation of oxide quality and its properties. We point out to opportunities usually not considered. There are, however, strong interplays between different processes and we cannot offer any ready recipe. More understanding is needed concerning the role of disorder and we are currently finishing a full study of oxidising species in amorphous oxides.

ACKNOWLEDGEMENTS

This work was supported in part by FECIT and by Fijutsu Laboratories, Japan. In particular we are grateful to Dr. Ross Nobes and Dr. Chioko Kaneta.

REFERENCES

1. Deal BE, Grow AS. General relationship for the thermal oxidation of silicon. J Appl Phys 1965;36:3770-8.
2. Sofield Cl, Stoneham AM. Oxidation of silicon: the VLSI gate dielectric?. Semicond Sd Tech 1995;10:215-44.
3. Coflot P, Gautherin G. Agius B, Rigo S. Rocket F. Low-pressure oxidation of silicon stimulated by low -energy electron bombardment. Phil Mag B 1985;52:1051-67.
4. Martinet C, Devine RAB. Brunel M. Oxidation of crystalline Si in an 02 plasma: growth kinetics and oxide characterisation. J Appl Phys 1997;81:6996-7005. 15) Stoneham AM. Taster PW. Image charges and their influence on the growth and the nature of thin oxide films. Phil Mag B 1987:55:237-52.
5. Wolters DR, Zegers-van Duijnhoven ATA. Advanced modeling of silicon oxidation. Microekctron Reliab 1998:38:259-64.
6. Jorgenson Pl. Effects of an electric field on silicon oxidation. 1 Chem Phys 1962;37:874-7.
7. Mills TG. Kroger FA. Electrical conduction at elevated temperatures in thermally grown silicon dioxide films. J Electrochem Soc 1973;120:1582-6.

8. Srivastava JK, Prasad M. Wagner Jr. JB. Electrical conductivity of silicon dioxide thermally grown on silicon. J Electrochem Soc 1985;132:955-63.
9. Xu J, Choyke WI, Yates JT. Enhanced silicon oxide film growth on Si (100) using electron impact. J Appl Phys 1997;82:6289-92.
10. II. Kazor A, Boyd 1W. Growth and modeling of cw-UV induced oxidation of silicon. 1 Appl Phys 1994:75:227-31.
11. Shamir N, Mihaychuk 1M. van Driel KM. Universal mechanism for gas adsorption and electron trapping on oxidized silicon. Phys Rev Lett 1999;82:359-61.
12. Kinon NU. Uren MI. Noise in solid.state microstructures - a new perspective on individual defects, interface states and low frequency (IfF) noise. Adv Phys 1989:38:367-468.
13. 11. Rochet F. Rigo S. Frown' M. SAnterroches C. Maillot C Roulet H. Durfour G. The thermal-oxidation of silken - the special case of the growth of very Adv Phys 1986:35:237-74.
14. Perdew JP. Electronic structure in solids. In: Zeisehe P. Eschrig H. editors. Berlin: Academie 1991.
15. Perdew JP. Chevary IA. Vosko SH. Jackson KA. Pederson MR. Fiolbais C. Atoms. molecules solids. and surfaces: applications of the generalized gradient approximation for exchange and correlation. Phys Rev B 1992:46:6671-87.
16. Kittle G. Furthmulkr J. Efficient iterative schemes for atfrinitio totalenergy calculations using plane•wave basis set. Phys Rev B 1996:54:11169-86.
17. Kresse G. Furthmulla J. Efficiency of ab•initio total energy calculations foe metals and semiconductors using a plane-wave basis WI. Comp Mater Sci 19964:15-50.
18. Makov V. Payne MC. Periodic boundary conditions in ab initio calculations. Phys Rev B 1995:51:4014-22.
19. Nlihaychuk IG. Shamir N. van Driel HM. Multiphoton photoemission and electric-field-induozd optical second-harmonic generation as probes of charge transfer across the Si/S102 interface. Phys Rev B 1999:592164-73.
20. May 1L. Hirose NI. The valence band alignment at ultrathin SiOs/Si interfaces. J Appl Phys 1997:81:1606-8.
21. Szymanski MA. Stoneham AM. Shluger AL. The roles of charged and neutral oxidising specks in silicon oxidation from a•initio calculations. Microelectronics Reliability 2000:40:567-70.
22. Fowler WB, Rudra JK, Zvanut ME, Feigl FL Hysteresis and Franck-Condon relaxation in insulator•emiamductor tunneling. Phys Rev B 1990:41:8313-7.
23. 21. Stoneham AM. Szymanski MA. Shluger AL. Dynamics of Silicon Oxidation. MRS Symp Proc 2000:592:3-14.

24. Mott NF. Rigo S. Rochet F. Stoneham AM. Oxidation of silicon. Phil Mag B 1989:03:189-212.
25. Stoneham ANL Szymanski MA. Shluger AL. Atomic and Ionic Processes of Silken Oxidation. Submitted to Phys Rev Lett.
26. Kurokawa A. Nakamura K. Ichimura S. Reduction of the interfacial Si displacement of ulrtathin Si02 on Si (100) formed by atmospheric pressure ozone. Appl Phys Lett 2000:76:493-S.

CITATION

M.A Szymanski, A.M Stoneham, A Shluger, The different roles of charged and neutral atomic and molecular oxidising species in silicon oxidation from ab initio calculations, Solid-State Electronics, Volume 45, Issue 8, August 2001, Pages 1233-1240, ISSN 0038-1101, http://dx.doi.org/10.1016/S0038-1101(00)00263-X.

CHAPTER 9

Strained Mosfets on Ordered Sige Dots

Johann Cervenka[1], Hans Kosina[1], Siegfried Selberherr[1], Jianjun Zhang[2], Nina Hrauda[2], Julian Stangl[2], Guenther Bauer[2], Guglielmo Vastola[3], Anna Marzegalli[3], Francesco Montalenti[3], Leo Miglio[3]

[1] Institute for Microelectronics, Technische Universität Wien, Gusshausstraße 27-29, 1040 Wien, Austria
[2] Institute of Semiconductor and Solid State Physics, Johannes Kepler University, Altenbergerstraße 69, 4040 Linz, Austria
[3] Department of Materials Science, University of Milano-Bicocca, Via Roberto Cozzi 53, 20125 Milano, Italy

ABSTRACT

The potential of strained DOTFET technology is demonstrated. This technology uses a SiGe island as a stressor for a Si capping layer, into which the transistor channel is integrated. The structure information of fabricated samples is extracted from atomic force microscopy (AFM) measurements. Strain on the upper surface of a 30 nm thick Si layer is in the range of 0.7%, as supported by finite element calculations. The Ge content in the SiGe island is 30% on average, showing an increase towards the top of the island. Based on the extracted structure information, three-dimensional strain profiles are calculated and device simulations are performed. Up to 15% enhancement of the NMOS saturation current is predicted.

INTRODUCTION

Strain engineering as a means to enhance electronic device performance has become an integral part of contemporary CMOS technology. In addition, novel device architectures have been proposed to improve the way in which strain is induced in the device channel. Schmidt and Eberl suggested to use self-assembled SiGe islands as stressors for Si capping layers [1]. Thereby higher strain values can be reached as compared to strained Si grown pseudomorphically on relaxed SiGe buffers.

In this work, the process for growing self-organized SiGe islands is briefly described, followed by an experimental and theoretical assessment of the strain into the capping layer, and a prediction of the performance enhancement of n-type FETs integrated in the strained capping layer by means of three-dimensional device simulation.

GROWTH OF REGULAR ARRAYS OF SIGE ISLANDS

The growth procedures are described in detail in [2] and [3]. The samples were grown on a Si(0 0 1) substrate, on which a square pattern of pits with a period of 800 nm has been defined by e-beam lithography, and transferred by reactive ion etching to form pits with a width of 170 nm and a depth of 75 nm. A 36 nm thick Si buffer layer was deposited on the pit-patterned substrates by molecular beam epitaxy, while the substrate temperature was ramped from 450 °C to 550 °C. Then the substrates were heated to a growth temperature of 720 °C, at which six mono-layers of Ge were deposited to form one dome-shaped SiGe island per pit. A Si capping layer of 30 nm thickness was deposited after cooling the substrate to 360 °C in order to avoid intermixing between the SiGe island and the Si cap. The surface morphology was investigated after each growth step by atomic force microscopy (AFM). Fig. 1 shows the AFM image of the final surface of the Si cap, which actually conformally replicates the surface after formation of the SiGe islands. Line scans across several pits and across a single pit are shown in Fig. 2, crossing the center of the pits and buried SiGe islands.

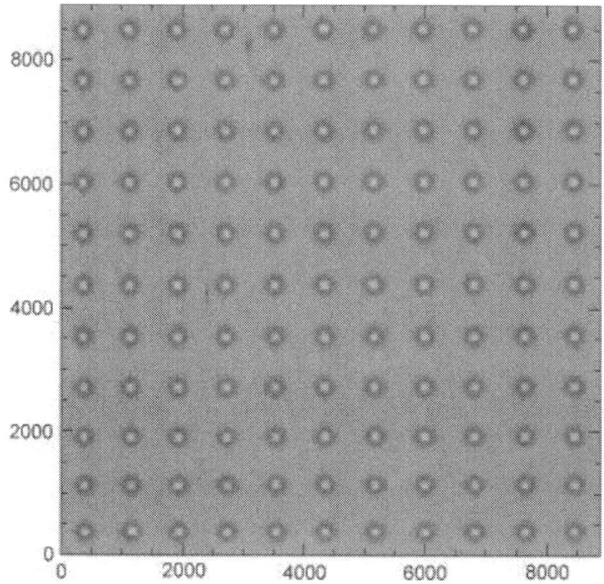

Figure 1. AFM micrograph of the sample surface after 30 nm of Si cap has been deposited onto SiGe islands grown in pits of a prepatterned Si substrate.

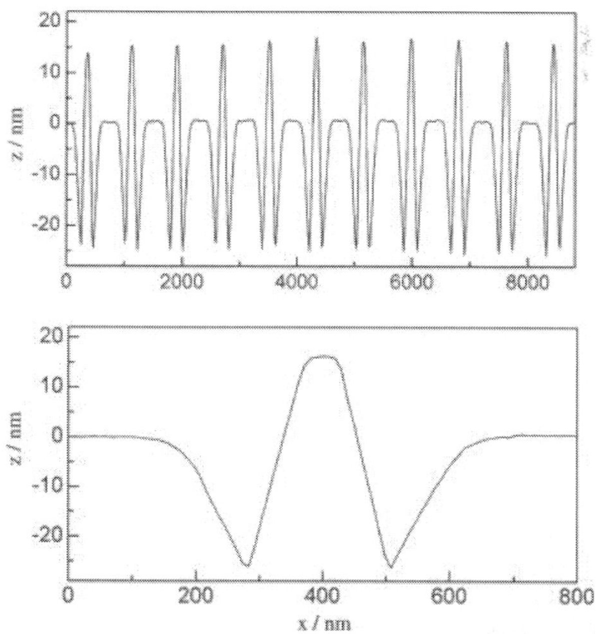

Figure 2. AFM linescan across the islands indicates the excellent homogeneity of the island array (upper panel). The shape of the cap is shown with high resolution in the lower panel.

MODELING OF THE STRAIN FIELD IN THE DOTFET PROCESS

The processes of substrate patterning and SiGe island growth have been optimized with respect to subsequent device fabrication. Excellent island ordering, as well as island shape, size, and composition uniformity have been achieved. This optimization has been supported by simulations. Three-dimensional AFM data have been directly imported into a finite element code for strain calculation. In particular, AFM data were taken both after the Si buffer growth (surface of the pit) and after Ge depositions. As compared to the ideal equilibrium shape of the islands as reconstructed from facet plots [4], the procedure used here includes more details of the actual structure such as edge rounding and the trench surrounding the island. In the elastic field calculations a Ge content profile in the island has been taken into account with an average Ge content of 0.3. In an iterative procedure the elastic energy is minimized. At the top of the island a higher Ge content is found, whereas at the lower interface the Ge content is reduced by intermixing with the Si buffer layer. The Si capping layer is deposited at sufficiently low temperature to prevent any significant intermixing of the deposited Si with the SiGe island, so that Ge content profiles can be estimated prior to capping. Then in the whole structure including the capping layer the strain field is calculated using a finite-element code. The sample considered here shows up to 0.7% biaxial, tensile strain at the upper surface of the 30 nm Si capping layer, where the transistor channel will be integrated (Fig. 3, Fig. 4 and Fig. 5). This value has also been confirmed by high-resolution X-ray diffraction [5] and Raman spectroscopy in conjunction with simulations [6].

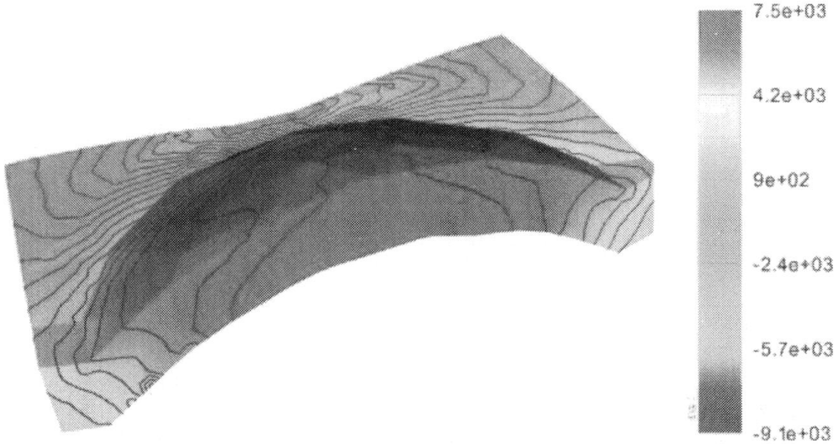

Figure 3. Strain component *exx* (channel length direction) in the Si capping layer.

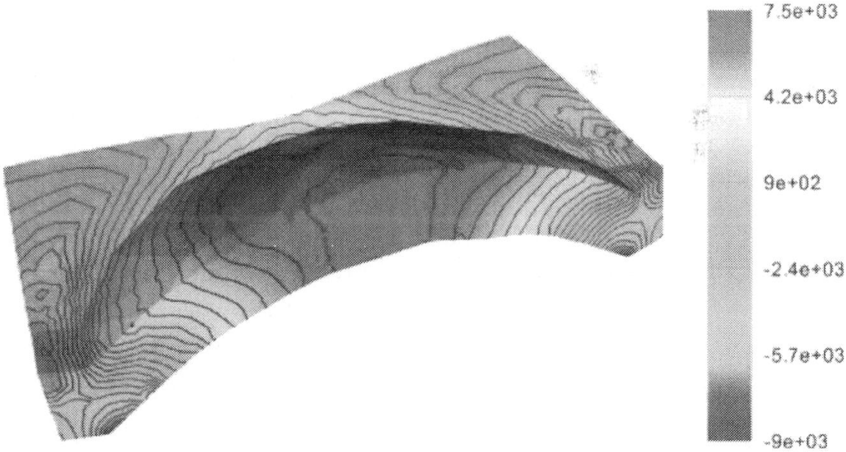

Figure 4. Strain component *eyy* (channel width direction) in the Si capping layer.

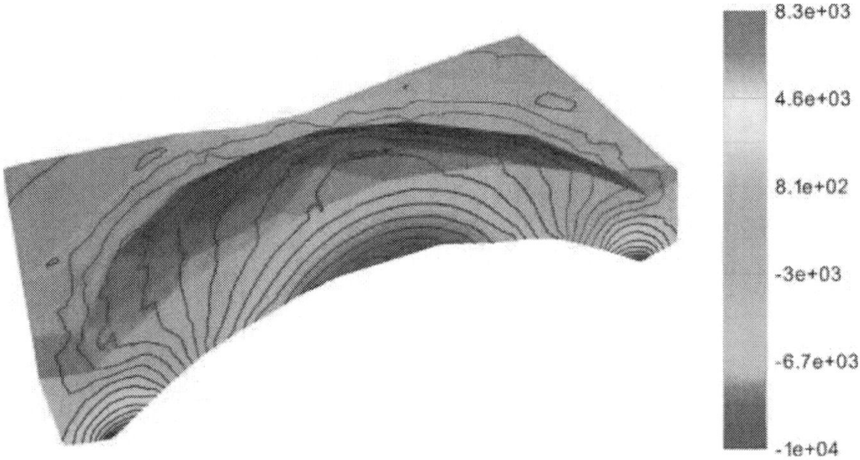

Figure 5. Strain component *ezz* (vertical direction) in the Si capping layer.

Removing the SiGe island at some stage of the process, a silicon on nothing (SON) device architecture can be realized. In this work, however, we consider a process in which the island is preserved. In this case the thermal budget must be kept sufficiently low to prevent intermixing of Si and Ge between the Si capping layer and the SiGe island. This can be achieved by state-of-the art low-temperature processing, including low-temperature formation of the gate stack and laser annealing of the source/drain implants. A low-complexity, dedicated n-channel MOSFET process has been developed [7].

THREE-DIMENSIONAL DEVICE SIMULATION

With the AFM data of the buffer and the SiGe island surfaces, the geometrical structure has been built. The Si capping layer is treated in the simulation by a conformal deposition. The correct representation of the Si cap is very important, since the simulated current is quite sensitive to the strain distribution at the surface. An unstructured mesh is used, which has to be sufficiently dense near the surface to resolve the surface inversion layer. On the other hand, in the SiGe island and the underlying Si buffer layer, lower point densities can be used.

Structure definition

The measured data is represented by z-values at defined xy-samples at the unstrained silicon surface, the surface of the dot, and the surface of the strained silicon regions. The tetrahedral mesher Netgen [8] is used for the three-dimensional mesh generation. This mesher features a robust meshing algorithm, which is able to handle the high aspect ratio of gate length/width to gate-oxide thickness. Input regions can be defined by a built-in solid modeler or by previously surface-meshed valid geometries. As the latter is not available in the measured data, a proper solid modeler geometry has to be built up from the existing data. Beside the usual primitives as tetrahedrons, bricks, and spheres, the solid modeler supports triangulated polyhedra. Therefore, several intermediate meshing stages have to be performed to achieve a proper input for the modeler.

In an initial step the xy-samples are two-dimensionally triangulated and the z-values are reassigned or interpolated to the delivered mesh points to achieve three-dimensional triangulated surfaces. As this mesh shows a huge number of surface elements, causing also a huge number of tetrahedral elements, an additional, adjustable smoothing stage was applied. Equipped with these data the solid modeler of Netgen was filled. To achieve a valid connectivity of the interfaces the following procedure is used (confer the final structure which is shown in Fig. 6 with the resulting mesh).

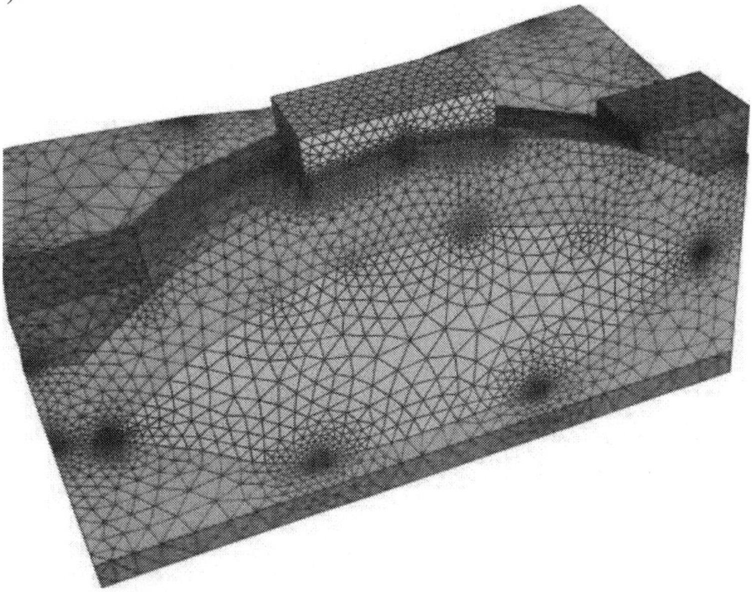

Figure 6. The geometry of the generated transistor structure. Only one half of the transistor is simulated.

First, the most interesting segment, the strained silicon is provided by a triangulated polyhedron of its top and bottom surface. The dot is built by an extruded lower surface of the dot-samples with subtraction of the strained silicon. The gate oxide is built as a thin extruded strained silicon surface and subtraction of strained silicon. All other segments, which are the gate, the bulk, the source and the drain region, are built as a cuboid with subtraction of the former segments. By the use of this solid modeler methodology, the interfaces between the segments are properly connected. Crucial for device simulation is the mesh density in the transistor channel. To achieve this desired high mesh density a maximum tetrahedron height for the mesh elements in the oxide region is assigned. Neighboring regions will start with these small elements growing towards the outer boundaries. With this technique an appropriate resolution in the channel region is achieved without boosting the overall number of tetrahedrons.

Characteristic data of the transistor are the gate stack consisting of a 1.5 nm thick oxide and a 60×60 nm^2 polysilicon gate, a bulk contact is attached to the Si buffer layer, source and drain regions are 60 nm wide and the implants are approximated by analytical profiles. In Fig. 6 the final structure with the simulation mesh is shown. The transistor is cut along its symmetry axis and only one half of the whole structure is analyzed.

In the Si cap layer a constant boron doping of 4×10^{18} cm^{-3} is assumed. Into the access regions to source and drain to the channel an arsenic profile with a maximum concentration of 5×10^{20} cm^{-3} is implanted (Fig. 7). Finally, the strain profile is interpolated from its originated ortho grid to the device simulation mesh. This mesh shows a high point density in the channel area. Therefore, the strain profile is well adapted in the active region.

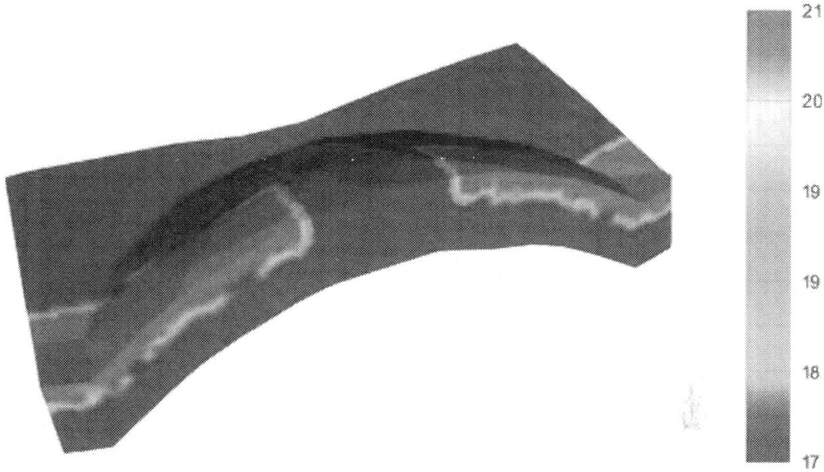

Figure 7. The arsenic doping with shown pn-junction in the capping layer.

Electrical characterization

Three-dimensional device simulations are performed using MINIMOS-NT [9]. The invoked physical models include the strain-dependent low-field mobility model described in [10] and the IMLDA quantum correction model [11]. To determine the strain-induced current enhancement, comparative simulations of the same structure are performed with and without taking the strain effect on the mobility into account.

In Fig. 8 the output characteristics for the unstrained and strained device are depicted. The transfer characteristics for drain voltages of 0.05 V and 1.5 V can be seen in Fig. 9.Fig. 10 shows the achieved drain current enhancement I_D/I_D^* (strained to unstrained drain current) as a function of the drain voltage. The current enhancement declines with growing gate voltage. Due to saturation of the electron velocity the enhancement also decreases towards higher drain voltages. At $V_{GS} = 1.5$ V and $V_{DS} = 1.5$ V an enhancement of 15% can be evaluated.

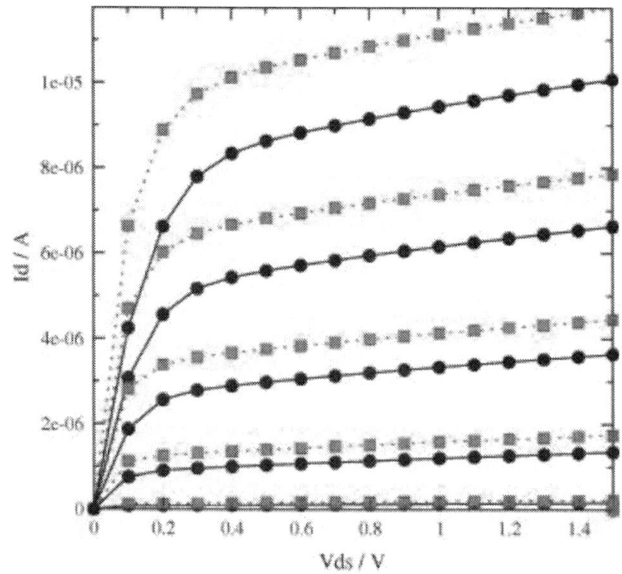

Figure 8. The unstrained (black bullets) and strained (red squares) output characteristics for gate voltages of 0.7 V, 0.9 V, 1.1 V, 1.3 V, and 1.5 V are shown. (For interpretation of the references to colour in this figure legend, the reader is referred to the web version of this article.)

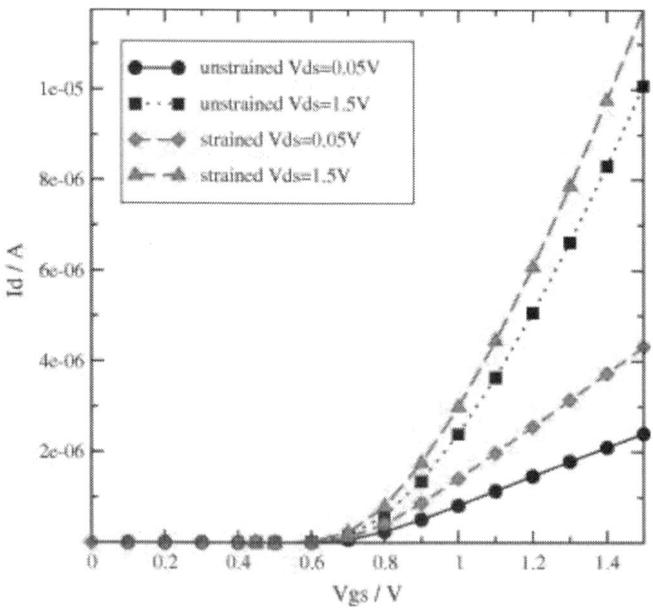

Figure 9. The unstrained and strained transfer characteristics for drain voltages of 0.05 V and 1.5 V.

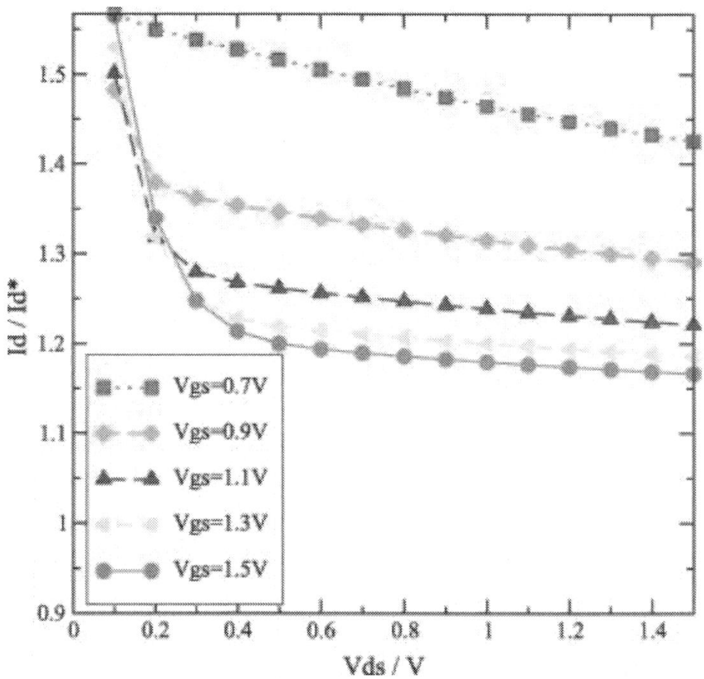

Figure 10. The current enhancement I_D/I_D^* for several gate voltages.

Three-dimensional impact of the dot

To examine the influence of the three-dimensional structure properties on the device characteristics the device has been cut along the channel axis. Depending on the position of the cut two-dimensional simulations were performed. On one hand side the stress profile varies towards lower stress values and on the other hand side the curvature and channel length decreases in the regions remote of the center plane. The result can be seen in Fig. 11. Here again the current enhancement I_D/I_D^* in dependence of the cut coordinate for V_{GS} = 1.5 V and V_{DS} = 1.5 V is plotted.

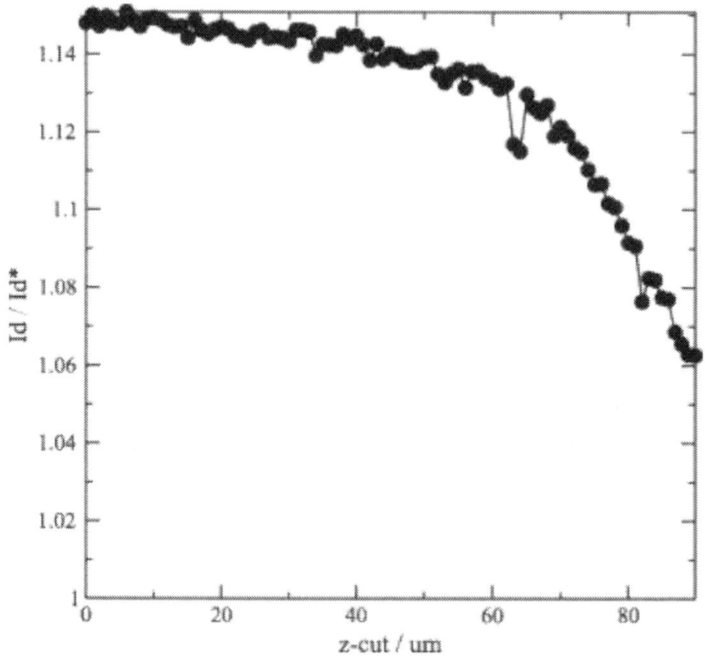

Figure 11. Current enhancement at a lateral cut of the original structure in dependence of the cut coordinate.

Comparison with an SOI structure

The DOTFET with removed SiGe dot has the electrostatic benefits of an SOI transistor. Due to the thermic problems arising in SOI technology this DOTFET structure is compared with an SOI structure. To have a comparable structure three different scenarios had been chosen, all based on the original device geometry, where different segments had been replaced by other materials:

- The original structure: consisting of the strained silicon bridge, the germanium dot, and the unstrained silicon buffer.
- The Silicon On Nothing structure: the germanium dot is etched away and (for simulation) replaced by air, which also can be produced by a modified process.
- The Silicon On Insulator structure: the unstrained silicon pit is replaced by insulator. This artificial geometry has been chosen to have

the same geometry of the active areas, the same doping, the same channel oxide, and the same gate influence.

If self heating is included, the thermal heat produced in the channel region of the original structure can be diverted through the silicon bridge, the germanium dot and the unstrained silicon region. Even with the dot removed the heat is diverted relatively good through the silicon bridge and the unstrained silicon region. However, for the SOI structure the thermal conductivity of the silicon buffer layer disappears. The heat has to be diverted through the contacts and surrounding insulators. The results of the simulation at $V_{GS} = 1.5$ V and $V_{DS} = 1.5$ V can be seen in Fig. 12, Fig. 13 and Fig. 14. Removing the dot only marginally changes the temperature profile. Only in air the profile varies, but in the active regions the temperature peak is nearly unaffected. Without the unstrained segment a dominant increase of the temperature in the channel region can be observed. This effect will also have significant influence on the mobilities and on impact ionization.

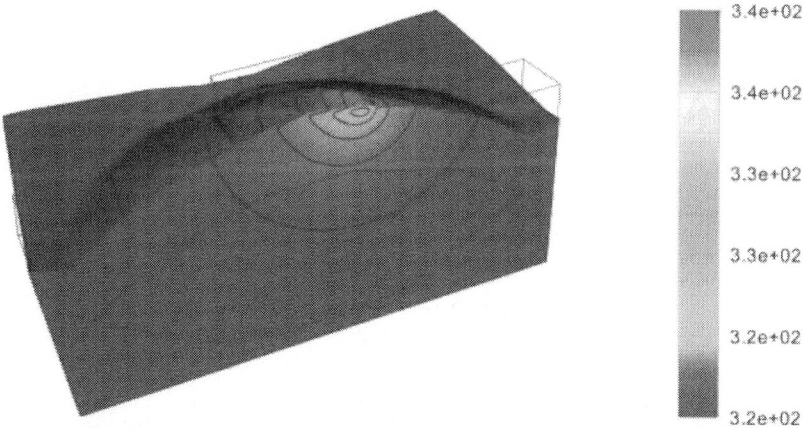

Figure 12. Temperature distribution at the original DOTFET structure with the germanium dot.

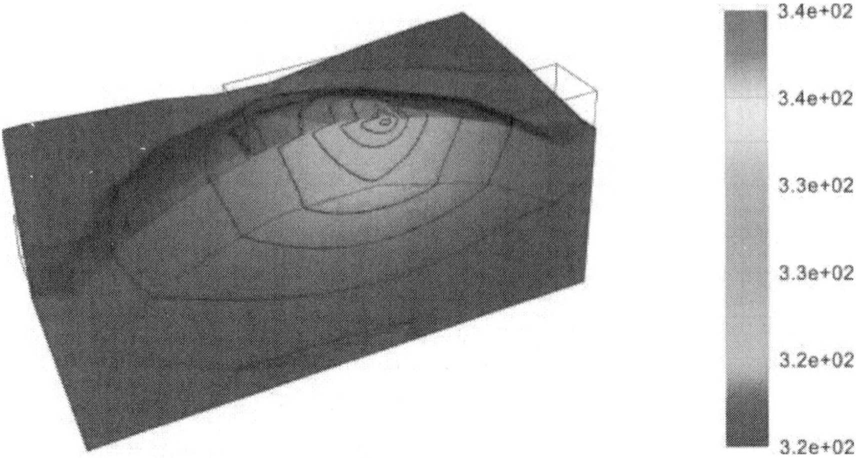

Figure 13. Temperature distribution at the SON device with removed germanium dot.

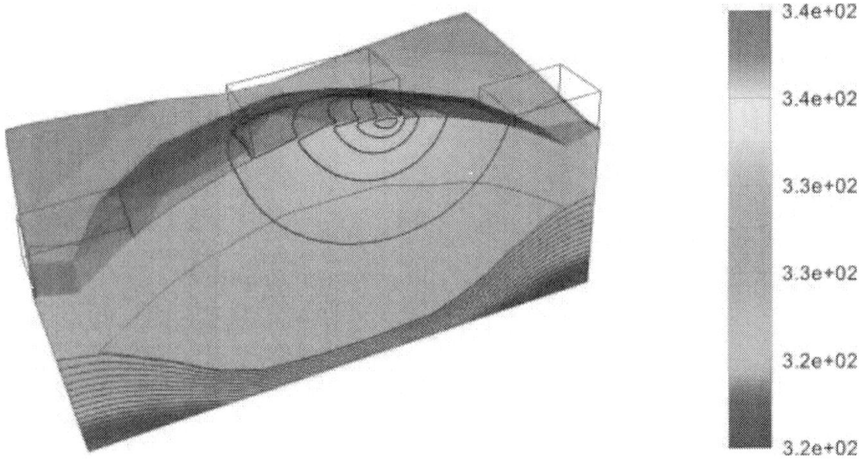

Figure 14. Temperature distribution at the SOI device with oxide instead of the unstrained silicon region.

DISCUSSION

In this section we compare the DOTFET structure with the so-called *Reverse Embedded SiGe structure* proposed by IBM in 2006 [12]. In the latter structure, strain is induced by elastic relaxation of a Si/SiGe bilayer. Source and drain are grown by selective epitaxy. We performed a strain calculation of the IBM structure and found a rather non-uniform strain profile with the strain maximum located in the middle of the channel. In [12] a uniaxial strain of 0.24% in the Si channel and a related drive current improvement of 15% are reported.

In comparison, the DOTFET process induces a more uniform strain up to 0.7% in the Si channel. The DOTFET structure is integrated in the coherently grown Si capping layer. Our simulations predict for the DOTFET the same drive current enhancement as has been reported for the IBM process, despite the strain in the latter is about three times lower (0.24% versus 0.7%). This can be partly attributed to different extraction methods for the current enhancement used in [12] and in our simulations. This comparison also indicates that the parameters used in our simulations give a conservative estimate of the current enhancement of the DOTFET.

CONCLUSION

In this work the potential of the DOTFET has been studied by three-dimensional simulations. The geometry has been extracted from actually fabricated samples. The strain field is obtained by comprehensive simulations verified by strain measurements on the strained Si layer. A conservative estimate for the NMOS drive current enhancement of about 15% is obtained.

ACKNOWLEDGMENT

This work has been supported by the EC, project No. 012150-2 (d-DOTFET) and the Austrian Science Fund FWF, projects F2507 and F2509 in the SFB 025 (IR-ON). We thank J. Moers and D. Grützmacher (FZ Jülich, Germany) for supplying the patterned Si wafers.

REFERENCES

1. Schmidt O, Eberl K. Self-assembled Ge/Si dots for faster field-effect transistors. IEEE Trans Electron Dev 2001;48(6):1175–9.
2. Zhong Z, Bauer G. Site-controlled and size-homogeneous Ge islands on prepatterned Si(0 0 1) substrates. Appl Phys Lett 2004;84(1):1922–4.
3. Zhang J, Stoffel M, Rastelli A, Schmidt O, Jovanovic V, Nanver L, et al. SiGe growth on patterned Si(0 0 1) substrates: surface evolution and evidence of modified island coarsening. Appl Phys Lett 2007;91:173115.
4. Stoffel M, Rastelli A, Tersoff J, Merdzhanova T, Schmidt O. Local equilibrium and global relaxation of strained SiGe/Si(00 1) layers. Phys Rev B 2006;74:155326.
5. Hrauda N, Zhang J, Stangl J, Rehman-Khan A, Bauer G, Stoffel M, et al. X-ray investigation of buried SiGe islands for devices with strain-enhanced mobility. J Vac Sci Technol B 2009;27(2):912–8.
6. Bonera E, Pezzoli F, Picco A, Vastola G, Stoffel M, Grilli E, et al. Strain in a single ultrathin silicon layer on top of SiGe islands:Raman spectroscopy and simulations. Phys Rev B 2009;79:075321.
7. Nanver L, Biasotto C, Jovanovic V, Moers J, Grützmacher D, Zhang J, et al.'SiGe dots as stressor material for strained Si devices. In: Proceedings of 5th international SiGe technology and device meeting (ISTDM), May 2010.
8. Schöberl J. NETGEN – automatic mesh generator.
9. MINIMOS-NT 2.1 User's Guide, Institut für mikroelektronik. Austria: Technische Universität Wien; 2004.
10. Dhar S, Kosina H, Palankovski V, Ungersböck E, Selberherr S. Electron mobility model for strained-Si devices. IEEE Trans Electron Dev 2005;52(4):527–33.
11. Jungemann C, Nguyen C, Neinhüs B, Decker S, Meinerzhagen B. Improved modified local density approximation for modeling of size quantization in NMOSFETs. In: Proceedings of international conference on modeling and simulation of microsystems 2001, vol. 1; 2001. p. 458–61.
12. Donaton R, Chidambarrao D, Johnson J, Chang P, Liu Y, Henson W, et al. Design and fabrication of MOSFETs with a reverse embedded SiGe (rev. e-SiGe) structure. In: IEDM Technical Digest; 2006. p. 465–68.

CITATION

Johann Cervenka, Hans Kosina, Siegfried Selberherr, Jianjun Zhang, Nina Hrauda, Julian Stangl, Guenther Bauer, Guglielmo Vastola, Anna Marzegalli, Francesco Montalenti, Leo Miglio, Strained MOSFETs on ordered SiGe dots, Solid-State Electronics, Volumes 65–66, November–December 2011, Pages 81-87, ISSN 0038-1101, http://dx.doi.org/10.1016/j.sse.2011.06.041.

CHAPTER 10

Impact of Oxidation and Reduction Annealing on the Electrical Properties of Ge/La2o3/Zro2 Gate Stacks

Christoph Henkel[1], *Per-Erik Hellström1*, *Mikael Östling*[1], *Michael Stöger-Pollach*[2], *Ole Bethge*[3], *Emmerich Bertagnolli*[3]

[1] Integrated Devices and Circuits, KTH Royal Institute of Technology, School of ICT, Stockholm, Sweden

[2] University Service Center for TEM, Vienna University of Technology, Vienna, Austria

[3] Institute for Solid State Electronics, Vienna University of Technology, Vienna, Austria

ABSTRACT

The paper addresses the passivation of Germanium surfaces by using layered La_2O_3/ZrO_2 high-*k* dielectrics deposited by Atomic Layer Deposition to be applied in Ge-based MOSFET devices. Improved electrical properties of these multilayered gate stacks exposed to oxidizing and reducing ambient during thermal post treatment in presence of thin Pt cap layers are demonstrated. The results suggest the formation of thin intermixed $La_xGe_yO_z$ interfacial layers with thicknesses controllable by oxidation time. This formation is further investigated by XPS, EDX/EELS and TEM analysis. An additional reduction annealing treatment further improves the electrical properties of the gate dielectrics in contact with the Ge substrate. As a result low interface trap densities on (1 0 0) Ge down to 3×10^{11} eV^{-1} cm^{-2} are demonstrated. The formation of the high-*k*$La_xGe_yO_z$ layer is in agreement with the oxide densification theory and may explain the improved I. The scaling potential of the respective layered gate dielectrics used in Ge-based MOS-based device structures to EOT of 1.2 nm or below is discussed. A trade-off between improved interface trap density and a lowered equivalent oxide thickness is found.

HIGHLIGHTS

- Ge surface passivation by scalable multilayer of La_2O_3/ZrO_2.
- $La_xGe_yO_z$ interfacial layers thickness controllable by oxidation time.
- Forming gas annealing improves D_{it} down to 3×10^{11} eV^{-1} cm^{-2} in presence of $La_xGe_yO_z$ interlayer.
- Trade-off between interface trap density and equivalent oxide thickness.

INTRODUCTION

Germanium (Ge) is a candidate to be integrated as a high mobility channel material in the complementary metal–oxide–semiconductor (CMOS) process flow at or below technology nodes of 16 nm [1] and [2]. One key reason for an increased scientific interest is the four times increased hole and two times increased electron mobility in bulk Ge compared to Si substrates [1]. Additionally, increased hole and electron mobility are reported in strained Ge substrates [3]. Until recently the lack of a suitable interface passivation methodology has hindered the possible application of Ge substrates in scaled CMOSFET device technologies, relying on the use of scaled high-*k* dielectric gate stacks [2] and [4]. However, in recent years excellent device performance was shown by improving the passivation properties of high-*k*/Ge interfaces [4], [5] and [6]. This was either achieved by the formation of ultrathin interfacial layers of Si/SiO_2 on top of the Ge p-channel, resulting in improved high-*k*/channel interface quality [5] or by introducing a thin interfacial layer of GeO_2 or GeON [4] and [6]. Only recently work on La-based rare-earth passivation of Ge surfaces is shown to yield improved device quality. The high band gap of La_2O_3 of 5.5–6.0 eV [7] and [8], combined with a conduction band offset of ~2.6 eV [8] and a *k*-value of 24–30 [9] make it a promising candidate to be integrated into scaled MOSFET devices. The required scaling potential of rare earth based metal–oxide–semiconductor (MOS) capacitor structures [10] and [11] and MOS field-effect-transistor (MOSFET) devices with equivalent oxide thickness (EOT) below 1 nm[12] and [13] was shown. However, in case of p-MOSFET devices still a lower mobility compared to thick GeO_2 based devices is reported. The achieved hole mobilities of 70–200 cm^2/V s [12], [14], [15], [16] and [17] are significantly lower than best reported values for thick GeO_2/Ge substrates with 575 cm^2/V s [4]. However, as the thickness of the GeO_2 reduces the hole mobility reduces [18]. These results directly connect the improvement in Ge surface passivation with the improvement in MOSFET mobility and on-current.

In earlier works it was shown that a combination of lanthanum-oxide (La_2O_3) interfacial layers formed by Atomic Layer Deposition (ALD) capped by a thin ALD zirconium-oxide (ZrO_2) high-k dielectric layer can yield to a good surface passivation of the Ge surface in combination with a high overall gate dielectric constant $k \sim 21$ and EOT below 1 nm [13]. Additionally, it was shown that the presence of thin platinum (Pt) layers deposited on top of La_2O_3/ZrO_2 gate dielectric yields a decrease of the interface trap density down to the mid-10^{11} eV^{-1} cm^{-2} regime by applying an oxygen annealing step [19]. By means of XPS it was concluded that a thin oxygen enriched interfacial layer was formed at the ALD La_2O_3 to Ge-channel interface.

In this work we present a detailed analysis of thermal post treatments of La_2O_3/ZrO_2 gate stacks in oxidative and reductive atmosphere and give a pathway for a possible use of the processed gate dielectrics in Ge-based MOSFET devices. By doing so we are able to form a low defect density high-k/Ge interface in combination with the reported scalability of the presented approach.

The results suggest the formation of a $La_xGe_yO_z$ interfacial layer due to the supply of excessive oxygen during the oxidation treatment. The experiments also show that the thickness of this interfacial layer can be effectively controlled by tuning the oxygen annealing time for a given oxygen pressure and temperature. Furthermore, it is shown that an additional reduction annealing improves the electrical performance of the device structure without changing the EOT. The improvement of interfacial trap density provides a direct link of the formation of a La-based interfacial layer with excellent passivation properties with regards to the Ge interface and an enhancement of the electrical performance of MOS based device structures.

The paper is arranged as follows; in the experimental Section 2 the process flow for the formation of thin capacitor structures by deposition of La_2O_3 and ZrO_2 dielectrics by ALD on (1 0 0) Ge surfaces is discussed. The different oxidizing and reducing post deposition annealing treatments are described. In the result Section 3 the findings by capacitance–voltage (*C–V*) and conductance–frequency (*G–f*) analysis for the given oxidizing and reducing annealing treatments performed are summarized. The interface trap density is determined by the conductance method [20] from the peak of conductance divided by the frequency. In addition, these results are in conjunction with X-ray photoelectron spectroscopy (XPS), transmission electron microscopy (TEM), energy-dispersive X-ray spectroscopy (EDX) line scans and electron energy loss spectroscopy (EELS) analysis performed on the gate stacks exposed to different annealing treatments. In Section 4possible mechanisms for the improved interface quality during the annealing treatments are discussed. Finally, the conclusion in Section 5 is correlating these results to their potential use in scaled Ge-based MOSFET devices used in future CMOS technology.

EXPERIMENTAL

Formation of gate stacks

Bulk (1 0 0) Ge *n*-type substrate with resistivity of 6–10 Ω cm were cleaned by cyclic treatment in deionized water and hydrofluoric acid (1.75%). Samples were dry blown using nitrogen and immediately transferred to the ALD reactor within 5 min. Subsequently, ALD was applied to deposit layers of 8 nm or 12 nm La_2O_3 capped by a layer of 1.5 nm ZrO_2. The Savannah 100 cross-flow reactor from Cambridge Nanotech was used. The precursors used were the La-precursor tris-(N,N′-diisopropylformamidinate)-lanthanum and the Zr-precursor tetrakis-(dimethylamino)-zirconium. Nitrogen was used as a purging gas. These precursors were kept at temperatures of 140 °C and 75 °C, respectively. As the oxygen supplying agent oxygen was used. Further information on the oxide deposition can be found in Refs. [13] and [19]. After the formation of the gate dielectric a thin Pt layer was deposited by sputter deposition of 5 nm Pt. From TEM images performed on cross-sections of the samples, the layer was found to be continuous. An annealing treatment in oxygen

(O_2) atmosphere was performed at atmospheric pressure at temperatures of 450 °C. In earlier experiments we showed that this Pt layer will assist the oxidation process and improve the electrical properties of the later Ge–high-*k* interface [19]. A ramp rate of 10 K/s was used for the temperature adjustment. Different annealing times were used from zero seconds (s) to 3600 s. Some samples were subjected to an annealing treatment in a reducing gas atmosphere using forming gas (N_2 (90%)/H_2 (10%)) for 30 min at 350 °C. Subsequently, Pt gate electrodes with thickness of 120 nm were deposited by PVD and structured by Ar sputter etching. Circular gate contact pads with 100 μm diameter were used. Pt back contacts were sputter deposited. A schematic cross section of the final MOS structure can be found in Fig. 1.

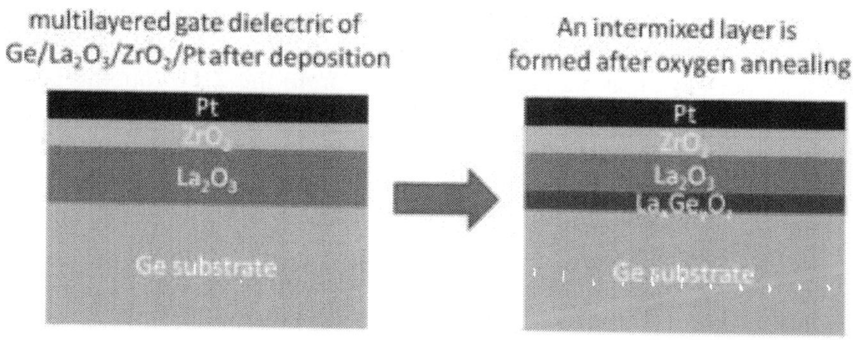

Figure 1. Left: Schematic of ALD La_2O_3/ZrO_2 multilayered gate dielectrics with a Pt gate electrode deposited on (1 0 0) Ge wafers. Right: the gate dielectric is capped with a 5 nm thin Pt metal and is subjected to an oxidation and FGA treatment. After the oxidation treatment a thicker $La_xGe_yO_z$ interfacial layer is formed in between the Ge substrate and the La_2O_3 gate dielectric.

Electrical and structural characterization of MOS capacitor structures

For the electrical characterization, *C–V* and *G–f* measurements were performed using Keithley's semiconductor characterization system 4200. For *C–V* measurements the dc-voltage is swept from −2 V in inversion to 2 V in accumulation as well as back for measurement of hysteresis. Equivalent oxide thickness was extracted using the Hauser CVC software [21]. The oxide thickness was determined by means of a Woolams α-SE spectroscopic ellipsometry setup. XPS analysis was performed using a Specs system. Depth resolved profiles were obtained by sputter-etching the gate stack surface using Ar-ions. [1] The XPS spectra were subsequently analyzed using CasaXPS software. For investigation of Ge oxidation states the Ge 2p peak position around 1217 eV was analyzed. The procedure is further described in Ref. [19]. EELS, EDX and transmission electron microcopy (TEM) analysis were performed using a TECNAI F20, FEI setup.

RESULTS

Capacitance–voltage characterization

First, the influence of the oxygen treatment on the process flow is investigated in terms of electrical properties of the gate dielectric stacks. Since it is known from earlier studies that a reaction of La_2O_3 with the Ge substrate occurs, the La_2O_3 layer thickness was intentionally selected to be relatively thick (8 nm) to allow an analysis of the structural composition of the interface region avoiding a contribution from the ZrO_2 capping layer. In a scaled MOSFET device using La_2O_3/ZrO_2 bilayer the La_2O_3 layer thickness should be decreased, as will be shown later. A gate stack consisting of 12 nm La_2O_3/1.5 nm ZrO_2 is subjected to different post deposition treatments in presence of a thin 5 nm Pt layer.

In Fig. 2a–c the influence of different annealing treatments using either O_2 annealing at 1 atm and 450 °C or N_2/H_2 annealing at 1 atm at 350 °C is compared. As a result the stretch out observed in samples of La_2O_3/ZrO_2 after only N_2/H_2-annealing treatment in the *C–V* curves vanishes even after very short oxygen annealing times of 20 s. From comparison of capacitance voltage characteristics obtained for longer oxygen annealing times a negative shift of flatband voltage is observed. Starting from the case of no oxygen annealing the flatband voltage is shifted to more negative values after a short anneal in oxygen for 20 s whereat a flatband voltage amounts to +0.24 V is measured. This value is further shifted to a negative value of −0.15 V after the oxidation treatment for 3600 s. A reduced humping near flatband voltage displays the decrease of the density of electrical active interface traps (D_{it}) near the Ge midgap. The gate voltage corresponding to the midgap position is obtained by fitting the respective CV curve to an ideal CV curve using the Hauser software [21]. Furthermore, a reduced hysteresis of ~100 mV is found after the oxygen annealing treatment. However, for longer annealing times a reduced oxide capacitance is found.

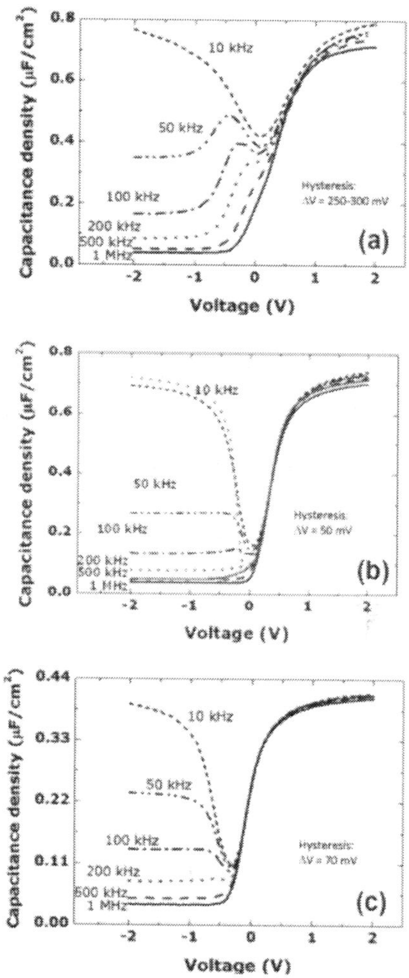

Figure 2. *C–V* characteristics of 12 nm La_2O_3 capped with 1.5 nm ZrO_2 deposited by ALD on (1 0 0) *n*-type Ge substrates. Before the post deposition annealing treatment a thin layer of 5 nm Pt is deposited by sputter deposition. (a) A reduction annealing in N_2/H_2 atmosphere at 350 °C for 30 min is applied; (b) an oxidation treatment at 1 atm at 450 °C for 20 s is applied before the FGA at 350 °C for 30 min; or; (c) an oxidation treatment at 1 atm at 450 °C for 3600 s is applied before the FGA at 350 °C for 30 min. After the different annealing treatments Pt gate electrodes are deposited and structured by sputter etching. The bump in (a) at ~−0.7 V is related to weak inversion response as pointed out in Ref. [31]. Note that the scale for graph (c) is different from graph (a) and (b). Simulated capacitance voltage characteristics for low (red, drawn line) and high (red, dotted line) measurement frequencies are additionally shown in (b). The curves are obtained by using Hauser CVC software [21] and fitting to experimental data for measurement frequencies of 500 kHz.

Interface trap density and equivalent oxide thickness

To further analyze the trade-off between the reduced D_{it} and the increased EOT of the presented layer structure of Ge/12 nm La_2O_3/1.5 nm ZrO_2, different oxide annealing times were applied starting from zero seconds up to 3600 s. The results of D_{it} for the different oxidizing and reducing annealing treatments are reported in Fig. 3. It can be clearly seen that the interface trap density is lowered with longer oxygen annealing times. The respective conductance frequency measurements are reported in Fig. 4 for the three shortest oxygen annealing times. Interestingly, this positive effect is even improved by performing the subsequent N_2/H_2 annealing treatment at 350 °C for 30 min in addition to the oxidizing annealing treatment. Here, an interface trap density down to ~3×10^{11} eV^{-1} cm^{-2} can be achieved. In addition, only little change in the EOT is found after the respective N_2/H_2 treatment (not shown) compared to samples not subjected to the FGA treatment after oxidation. In Fig. 5 the dependence of the measured equivalent oxide thickness during thermal oxidation treatment at 450 °C at 1 atm O_2 is shown. Starting with an EOT of 3.8 nm for the layer structure of 12 nm La_2O_3/1.5 nm ZrO_2 increased EOT's are found in the first 10–60 s of the oxygen annealing treatment from 0.5 to 0.8 nm. As a comparison the formation of GeO_2 on bare Ge substrates from thermal oxidation grown performed at atmospheric pressure at 450° is shown. A similar growth characteristic is found.

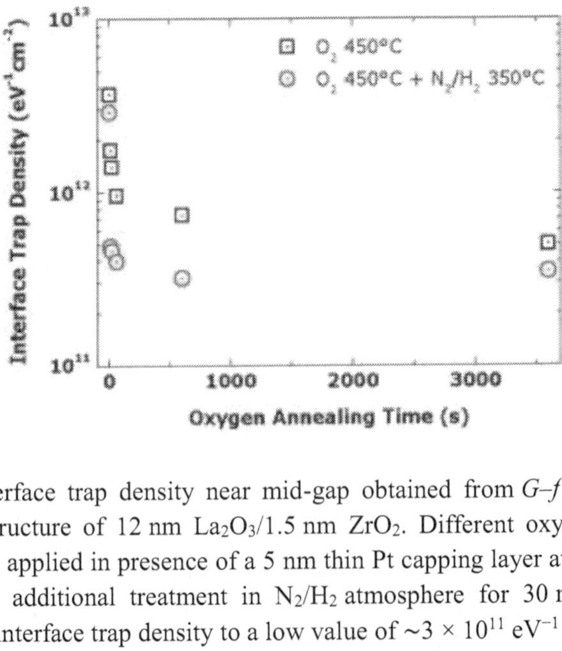

Figure 3. Interface trap density near mid-gap obtained from *G–f* measurements for a layer structure of 12 nm La_2O_3/1.5 nm ZrO_2. Different oxygen annealing treatments are applied in presence of a 5 nm thin Pt capping layer at a O_2 pressure of 1 atm. An additional treatment in N_2/H_2 atmosphere for 30 min at 350 °C improves the interface trap density to a low value of ~3×10^{11} eV^{-1} cm^{-2}.

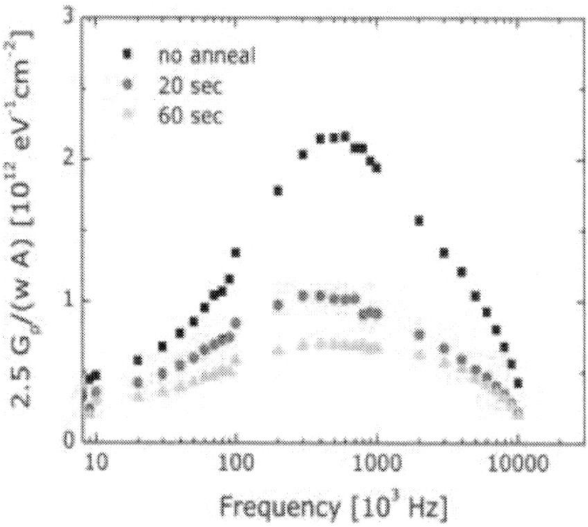

Figure 4. Typical conductance measurements are shown as a function of frequency for the layer structure of 12 nm La_2O_3/1.5 nm ZrO_2. The result for samples subjected to no oxygen anneal, a 20 s and a 60 s oxygen anneal are compared. The respective interface trap density was obtained from the peak of the measured conductance divided by the capacitor area and measurement frequency according to the conductance method [20].

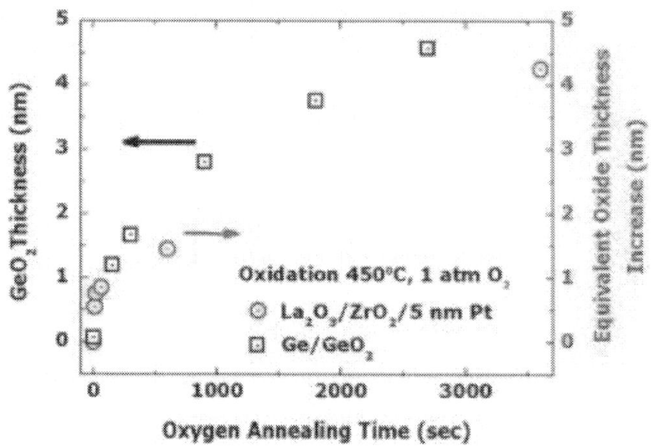

Figure 5. Increase in EOT as a function of oxygen annealing time at 1 atm at 450 °C for a layer of 12 nm La_2O_3/1.5 nm ZrO_2. The starting value of EOT is 3.8 nm. Nearly no change in the measured EOT after performing an additional reduction annealing step in N_2/H_2 is observed (not shown). Additionally, the oxidation of a clean Ge surface is shown. Here, the physical GeO_2 oxide thickness is shown. The surface is subjected to the same oxidation treatment at 450 °C at 1 atm.

Structural characterization

In order to investigate the structural properties of the gate dielectric to Ge interface different analysis methods were applied. TEM analysis was performed on stacks of 12 nm La_2O_3/1.5 nm ZrO_2 which were annealed for either 60 s or 3600 s in oxygen at 450 °C. Fig. 6 shows the respective results for the gate stacks exposed to the oxygen annealing treatment in presence of the thin Pt layer. Comparing the two gate stacks an increase in the interfacial layer thickness can be observed in case of the longer annealing time. The lighter contrast of this interlayer in TEM images also suggests the enrichment in lighter elements that would be expected in case of increased oxygen incorporation. This is also consistent with an oxygen densification in the interlayer as pointed out by Tsoutsou and coworkers [22]. From the combination of TEM analysis and CV measurements of the samples oxidized for either 60 s or 3600 s the dielectric constant of the interfacial layer can be extracted with $k = 8–9$. This is in good agreement with the reported value for $La_xGe_yO_z$ of 9–11 [23]. Assuming a change of 0.5 nm in EOT for oxygen annealing times for the case of 10 s O_2 annealing (as found from Fig. 5) an additional formation of a 1.1 nm $La_xGe_yO_z$ interfacial layer can be assumed. Interestingly, a strong surface roughening is present for long oxygen annealing times as can be seen from the interface of $La_xGe_yO_z$ with Ge.

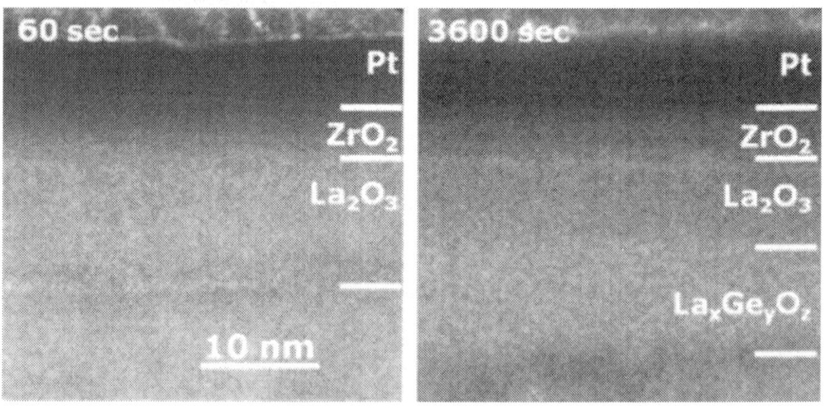

Figure 6. TEM analysis of 12 nm La_2O_3/1.5 nm ZrO_2 annealed for 60 s (left) or 3600 s (right) on O_2 1 atm at 450 °C in presence of 5 nm Pt capping layer. Three distinct layers on top of the Ge substrate can be identified, which are La_2O_3, a thin ZrO_2, and the thin Pt capping layer structure (dark contrast). In addition, the formation of a lighter contrast interfacial layer can be seen for long oxygen annealing treatment of 3600 s (right).

To analyze in more detail the formation of the interfacial layer at the La_2O_3 to Ge interface a further compositional analysis is performed on the shown TEM cross-sections. Fig. 7shows the results of depth resolved EELS analysis for the O profile in combination with EDX line scans for Ge, La, Pt and Zr performed at samples with 12 nm La_2O_3/1.5 nm ZrO_2 on Ge. These samples were also subjected to a 3600 s annealing in O_2 at 450 °C. As can be seen from the La and Ge peaks, at the interface between La_2O_3 and Ge an intermixed layer is formed. The data of the Ge signal indicates a Ge or GeO [24] in-diffusion into the La_2O_3 dielectric layer. These results are further supported by results from depth resolved XPS analysis applied for a stack of 8 nm La_2O_3/1.5 nm ZrO_2subjected to the same oxygen annealing treatment seen in Fig. 8. Here the line spectra for La 3d 5/2 was recorded in a range of 830–844 eV, O 1s in a range of 525–535 eV, Zr 3d in a range of 174–188 and Pt 4f in a range of 64–80 eV. Fig. 9 is additionally showing the narrow scan for these XPS analysis around the Ge 2p peak energy for different sputter times of the gate dielectric stack. The peak was recorded from 1214 to 1222 eV. As can be seen first a peak shifted by ~2.3 eV is observed, related to La–Ge–O bonding[25]. Finally, the Ge 2p signal is obtained related to the Ge substrate. This is in well agreement with a formation of a $La_xGe_yO_z$ interfacial layer.

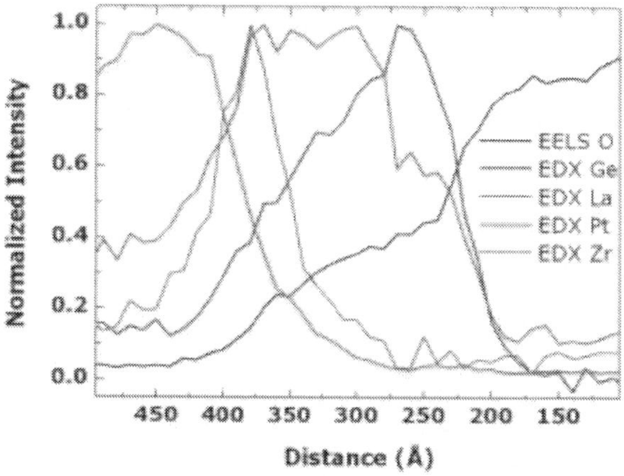

Figure 7. EELS and EDX line scan analysis of 12 nm La_2O_3/1.5 nm ZrO_2 after oxygen annealing treatment for 3600 s at 1 atm and 450 °C. An intermixed interfacial layer of Ge, La and O is observed at the La_2O_3/Ge interface.

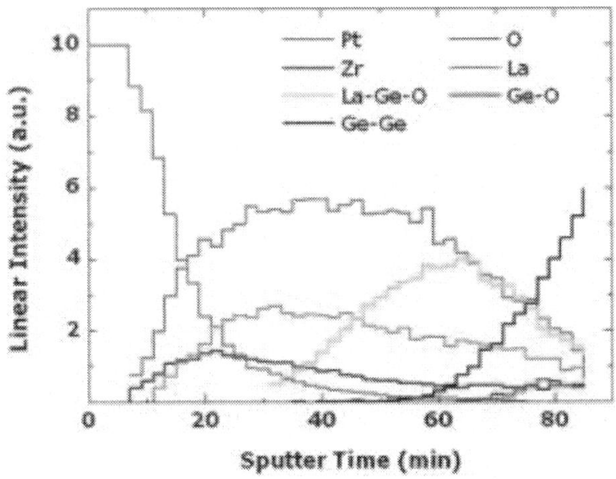

Figure 8. Depth resolved XPS analysis applied to a layer structure of 8 nm La_2O_3/1.5 nm ZrO_2 subjected to an oxygen annealing treatment of 3600 s at 450 °C and 1 atm. A clear signal of La–Ge–O signal can be found at the interface of Ge–La_2O_3 interface. Here, La–Ge–O denotes the Ge binding state shifted by 2.3 eV with respect to the Ge 2p peak position (Ge–Ge) and Ge–O the Ge^{2+} oxidation state shifted by 1.5 eV.

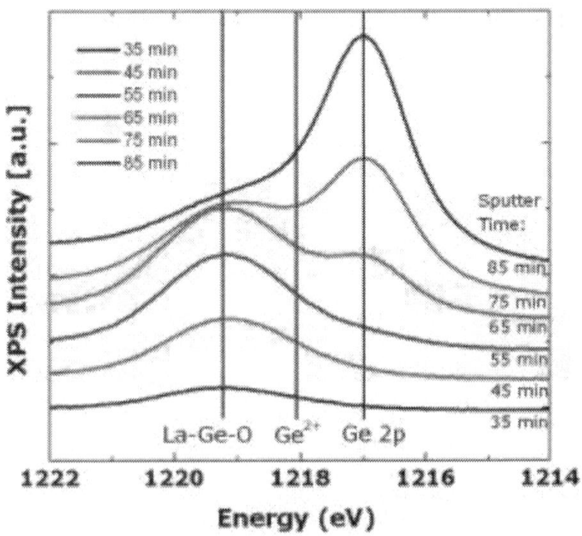

Figure 9. Narrow spectrum of Ge 2p XPS signal around Ge 2p peak position. The intensity of curves is shifted by a constant offset for the different sputter times shown.

DISCUSSION

In order to understand the improvement in measured electrical characteristics with the respective annealing treatment, first, the formation of the $La_xGe_yO_z$ interfacial layer shall be discussed. After the wet chemical processing for cleaning of the Ge substrates samples are subjected to cleanroom atmosphere and transferred to the ALD reaction chamber within several minutes. However, it is expected that a thin GeO_x suboxide forms in this processing scheme. It was pointed out by Houssa and coworkers that La_2O_3 in contact to the GeO_x results in the formation of only La–O–Ge bonds near the interface. The reason is the fourfold coordination of La atoms in the GeO_x matrix. Interestingly, the resulting band structure of Ge at the interface is free of surface states and thus, La_2O_3layers are predicted to show a good passivation of the Ge surface [23]. Further on, the samples are subjected to an oxide annealing treatment. Here, the driving force for the formation of $La_xGe_yO_z$ interfacial layers is suggested to be related to oxide densification as proposed by Tsoutsou and coworkers [22]. This theory proposes that the less dense oxide La_2O_3 with its strong electropositive La atom tries to attract as much oxygen atoms to its surrounding as possible, thereby reacting with the more oxygen dense GeO_2species [11]. The constant increased supply of oxygen during the oxygen annealing is thus leading to the observed growth of a thick $La_xGe_yO_z$ interlayer, instead of a formation of interfacial GeO_2. This is in agreement with the finding of an interfacial layer with a value of $k = 8–9$ which was obtained by comparison of TEM and $C–V$ analysis. As we report in [19] a thin Pt cap layer deposited before the oxidation annealing further assists the oxidation process, most likely by the formation of atomic oxygen by dissociating O_2molecules from the oxygen gas atmosphere.

No change in the equivalent oxide thickness is observed during the forming gas anneal. This suggests that hydrogen species leave the $La_xGe_yO_z$ layer thickness constant without any further layer growth. As a result lowest interface trap densities down to 3×10^{11} eV^{-1} cm^{-2} are achieved. Remarkably, in the case where the gate stacks are not subjected to an oxidation anneal before the FGA, little or no improvement of the interface trap density is observed, as can be seen from Fig. 3. This refers to the first data point, of 0 s oxygen anneal. For these samples we assume that no or only a thin $La_xGe_yO_z$interfacial layer is present. A similar improvement in interface trap density was observed for MBE La_2O_3 layers in contact to (1 0 0)-Ge after forming a La – germinate interfacial layer [15]. Our results for ALD La_2O_3/ZrO_2 are similar to results reported for ALD grown Al_2O_3 on interfacial GeO_2 with D_{it} of low to mid-10^{11} eV^{-1} cm^{-2} [26], high-pressure oxidized Ge-(1 1 1)/Y_2O_3 of ~10^{11} eV^{-1} cm^{-2} [27] and are well below reported values for Ge in direct contact to ZrO_2 or HfO_2 with >10^{12} eV^{-1} cm^{-2} [2].

Several studies address the passivation properties of high-*k* dielectrics in contact to Ge surfaces [28]. In general two types of concerns are addressed. These are either the diffusion of volatilization byproducts from annealing of Ge substrates such as Ge or GeO atoms [11] and [29] or the intrinsic defects related to the defective Ge surface [11] and [23]. It is known that by forming a $La_xGe_yO_z$ interlayer the formation of volatile GeO species is suppressed. Furthermore, as was found by combination of XPS and UPS, a $La_xGe_yO_z$ interlayer might improve the intrinsic defects present at the Ge surface, which are responsible for Fermi-Level pinning [11]. Since we observe the strongest improvement in interface trap densities already within short oxygen annealing times and thus, during the formation of a thin $La_xGe_yO_z$ interfacial layer in contact with Ge, the improved electrical passivation can be related to the $La_xGe_yO_z$ layer. The observed negative shift in flatband voltage could additionally be related to a lowered amount of interface charge density D_{it} during the oxygen anneal as discussed by Tsipas and Dimoulas [23]. However, this shift of threshold voltage might also be caused by an increased number of fixed charges. Addressing the performance of MOSFET devices, it is not clear if this improvement of interface trap density can be translated to improved device characteristics. No improved mobility in gate stacks of Al_2O_3/La_2O_3/Ge was observed by Rossel and coworkers [16]. However, other reports by Andersson and coworkers suggest a positive influence of FGA in both D_{it} and mobility of holes in p-channel devices using comparable gate dielectric stacks of HfO_2/La_2O_3/Ge [15].

For the presented oxygen annealing approach equivalent oxide thicknesses of 1.2–1.43 nm can be achieved in case of a 1 nm thin La_2O_3 interfacial layer capped by a thicker layer of 7 nm ZrO_2 as reported in Ref. [19]. We note that for such thin layers of La_2O_3 a $La_xGe_yO_z$ interlayer can be already observed after deposition of ZrO_2 by ALD [30]. In general, the thickness of the $La_xGe_yO_z$ interfacial layer with a lower k-value of 8–9 as shown in this study, should be optimized in order to find a good passivation of surface defects while keeping the equivalent oxide thickness low. Concerning this issue, our results suggest that a thin $La_xGe_yO_z$ interlayer with thickness of ~1.1 nm already formed after 10 s oxygen annealing provides a good surface passivation with density of interface states in the low 10^{11} eV^{-1} cm^{-2} range and a related additional trade-off in EOT of only 0.5 nm.

CONCLUSION

A pathway for improving the interface quality of ALD deposited La_2O_3 interface passivation layers on Ge substrates is given. These layers were capped with ZrO_2 and furthermore a Pt-assisted oxidation process in combination with a reduction annealing treatment in N_2/H_2 was applied. For a layered gate stack structure of Ge/La_2O_3/ZrO_2/Pt it was shown that interface trap densities down to $\sim 3 \times 10^{11}$ eV^{-1} cm^{-2} can be achieved in combination with minimal discount in EOT of 0.5 nm for short time annealing in oxygen atmosphere and a subsequent annealing step in reducing atmosphere. As the main driving force of this improvement the formation of a La–germanate interfacial layer was identified with dielectric constant of 8–9. An improved surface passivation was achieved already for thin $La_xGe_yO_z$ interfacial layers with thickness of 1–2 nm and minimal EOT offset. Thus a trade-off between improved surface quality and scalability of the presented approach was observed. However, the interlayer thickness was found to be adjustable by the time of an oxide annealing treatment. In a similar way the interlayer thickness may be adjustable by choosing a La_2O_3 starting thickness smaller than the value of 8 or 12 nm reported in this study. In summary, the presented results give evidence for an efficient approach to improve the electrical surface passivation required for future Ge-based MOSFET device technology. Additionally, the insertion of the thin Pt metal interlayer with a high vacuum work function may also offer advantages in Ge-pMOSFET device performance.

ACKNOWLEDGMENTS

This work is funded by the Austrian Science Fund (FWF), Project Number P 19787-N14. The Zentrum für Mikro-und Nanostrukturen, ZMNS, is gratefully acknowledged for support. The work is partly funded by ERC Advanced Grant OSIRIS.

REFERENCES

1. Heyns M, Tsai W. Ultimate scaling of CMOS logic devices with Ge and III–V materials. MRS Bull 2009;34:487–92.
2. Houssa M, Chagarov E, Kummel A. Surface defects and passivation of Ge and III–V interfaces. MRS Bull 2009;34:504.
3. Watling JR, Riddet C, Chan MKH, Asenov A. Simulation of hole mobility in doped relaxed and strained Ge layers. J Appl Phys 2010;108:093715.
4. Takagi S, Dissanayake S, Takenaka M. High mobility Ge-based CMOS device technologies. Key Eng Mater 2011;470:1–7.
5. Mitard J et al. Record ION/IOFF performance for 65 nm Ge pMOSFET and novel Si passivation scheme for improved EOT scalability. IEEE International Electronic Device Meeting. IEDM; 2008. p. 1.
6. Nishimura T, et al. Electron mobility in high-k Ge MISFETs goes up to higher. Symposium on VLSI technology digest of technical papers. 2010; 209.
7. Adachi G, Imanaka N. The binary rare earth oxides. Chem Rev 1998;98:1479–514.
8. Robertson J, Falabretti B. Band offsets of high-k gate oxides on high mobility semiconductors. Mater Sci Eng B 2006;135:267–71.
9. Engström O, Raeissi B, Hall S, Buiu O, Lemme MC, Gottlob HDB, et al. Navigation aids in the search for future high-k dielectrics: physical and electrical trends. Solid-State Electron 2007;51:622–6.
10. Dimoulas A, Tsoutsou D, Panayiotatos Y, Sotiropoulos A, Mavrou G, Galata SF, et al. The role of La surface chemistry in the passivation of Ge. Appl Phys Lett 2010;96:012902.

11. Dimoulas A, Tsoutsou D, Galata S, Panayiotatos Y, Mavrou G, Golias E. Ge surfaces and its passivation by rare earth lanthanum germanate dielectric. ECS Trans 2010;33:433.
12. Henkel C, Abermann S, Bethge O, Pozzovivo G, Klang P, Reiche M, et al. Ge pMOSFETs with scaled ALD La2O3/ZrO2 gate dielectrics. IEEE Trans Electron Dev 2010;57:3295.
13. Abermann S, Bethge O, Henkel C, Bertagnolli E. atomic layer deposition of ZrO2/La2O3 high-k dielectrics on Germanium reaching 0.5 nm equivalent oxide thickness. Appl Phys Lett 2009;94:262904.
14. Gu J, Liu Q, Xu M, Celler G, Gordon R, Ye P. High performance atomic-layerdeposited LaLuO3/Ge-on-insulator p-channel metal–oxide–semiconductor field-effect transistor with thermally grown GeO2 as interfacial layer. Appl Phys Lett 2010;97:012106.
15. Andersson C, Sousa M, Marchiori C, Webb D, Caimi D, Siegwart H, et al. Impact of La2O3 thickness on HfO2/La2O3/Ge capacitors and p-channel MOSFETs. Advanced Functional Materials. 2009; p. 0–3.
16. Rossel C, Dimoulas A, Tapponnier A, Caimi D, Webb DJ. Andersson C, et al. Ge p-channel MOSFETs with La2O3 and Al2O3 gate dielectrics. In: Proc. 38th Eur. solid-state device res. conf.2008. p. 70–82.
17. Dimoulas A, Panayiotatos Y, Sotiropoulos A, Tsipas P, Brunco DP, Nicholas G, et al. Germanium FETs and capacitors with rare earth CeO2HfO2 gates. Solid State Electron 2007;51:1508–14.
18. Caymax M, Eneman G, Bellenger F, Merckling C, Delabie A, Wang G, et al. Germanium for advanced CMOS anno 2009: a SWOT analysis. IEDM; 2009.
19. Henkel C, Bethge O, Abermann S, Puchner S, Hutter H, Bertagnolli E. Pt-assisted oxidation of (1 0 0)-Ge/high-k interfaces and improvement of their electrical quality. Appl Phys Lett 2010;97:152904.
20. Sze SM, Kwok KNG. Physics of semiconductor devices. 3rd Ed.. Hoboken (New Jersey): John-Wiley and Sons; 2007. p. 219–223.
21. Hauser JR, Ahmed K. Characterization of ultra-thin oxides using electrical C–V and I–V measurements. In: Characterization and metrology for ULSI technology. 1998. p. 235.
22. Tsoutsou D, Panayiotatos Y, Sotiropoulos A, Mavrou G, Golias E, Galata SF, et al. Chemical stability of lanthanum germanate passivating layer on Ge upon highk deposition: a photoemission study on the role of La in the interface chemistry. J Appl Phys 2010;108:064115.
23. Tsipas P, Dimoulas A. Modeling of negatively charged states at the Ge surface and interfaces. Appl Phys Lett 2009;94:012114.
24. Golias E, Tsetseris L, Dimoulas A, Pantelides ST. Ge volatilization products in high-k gate dielectrics. Microelectron Eng 2011;88:427–30.

25. Mavrou G, Tsipas P, Sotiropoulos A, Galata S, Panayiotatos Y, Dimoulas A, et al. Very high-j ZrO2 with La2O3 (LaGeOx) passivating interfacial layers on Germanium substrates. Appl Phys Lett 2008;93:212904.
26. Delabie A, Bellenger F, Houssa M, Conard T, Van Elshocht S, Caymax M, et al. Effective electrical passivation of Ge (100) for high-k gate dielectric layers using Germanium oxide. Appl Phys Lett 2007;91:082904.
27. Nishimura T, Lee CH, Wang SK, Tabata T, Kita K, Nagashio K, et al. Electron mobility in high-k Ge-MISFETs goes up to higher. In: Symposium on VLSI technology digest of technical papers. 2010. p. 209.
28. Houssa M, Pourtois G, Caymax M, Meuris M, Heyns M. First-principles study of the structural and electronic properties of (1 0 0)Ge/Ge(M)O2 interfaces (M = Al, La, or Hf). Appl Phys Lett 2008;92:242101.
29. Golias E, Tsetseris L, Dimoulas A, Pantelides ST. Ge volatilization products in high-k gate dielectrics. Microelectron Eng 2010;88:427.
30. Bethge O, Henkel C, Abermann S, Pozzovivo G, Stoeger-Pollach M, Werner W, et al. Stability of La2O3 and GeO2 passivated Ge surfaces during ALD of ZrO2 high-k dielectric. Appl Surface Sci 2012;285:3444.
31. Martens K, Chui CO, Brammerts G, De Jaeger B, Kuzum D, Meuris M, et al. On the correct extraction of interface trap density of MOS devices with highmobility semiconductor substrates. IEEE Trans Electron Dev 2007;55:547.

CITATION

Christoph Henkel, Per-Erik Hellström, Mikael Östling, Michael Stöger-Pollach, Ole Bethge, Emmerich Bertagnolli, Impact of oxidation and reduction annealing on the electrical properties of Ge/La2O3/ZrO2 gate stacks, Solid-State Electronics, Volume 74, August 2012, Pages 7-12, ISSN 0038-1101, http://dx.doi.org/ 10.1016/j.sse.2012.04.004.

CHAPTER 11

Automatic Generation Control with Thyristor Controlled Series Compensator Including Superconducting Magnetic Energy Storage Units

Saroj Padhan, Rabindra Kumar Sahu, Sidhartha Panda

Department of Electrical Engineering, Veer Surendra Sai University of Technology (VSSUT), Burla 768018, Odisha, India

ABSTRACT

In the present work, an attempt has been made to understand the dynamic performance of Automatic Generation Control (AGC) of multi-area multi-units thermal–thermal power system with the consideration of Reheat turbine, Generation Rate Constraint (GRC) and Time delay. Initially, the gains of the fuzzy PID controller are optimized using Differential Evolution (DE) algorithm. The superiority of DE is demonstrated by comparing the results with Genetic Algorithm (GA). After that performance of Thyristor Controlled Series Compensator (TCSC) has been investigated. Further, a TCSC is placed in the tie-line and Superconducting Magnetic Energy Storage (SMES) units are considered in both areas. Finally, sensitivity analysis is performed by varying the system parameters and operating load conditions from their nominal values. It is observed that the optimum gains of the proposed controller need not be reset even if the system is subjected to wide variation in loading condition and system parameters.

INTRODUCTION

Load Frequency Control (LFC) is a very important issue in modern power system operation and control for supplying sufficient and reliable electric power with good quality. The main goal of the LFC is to maintain the system frequency of each area and the tie line power within tolerable limits with variation in load demands [1]. For power balance, the power generated should match with the total load demanded and associated system losses. However the load demands fluctuate randomly causing a mismatch in the power balance and thereby deviations in the area frequencies and tie-line powers from their respective scheduled values, called Automatic Load Frequency Control (ALFC) [2] and [3]. Due to the complexity of the modern power system, superior intelligent control design is essential. Literature study reveals that several control strategies have been proposed by many researchers over the past decades for LFC of power system. Many control and optimization techniques such as classical, optimal, Genetic Algorithm (GA), Particle Swarm Optimization (PSO), Fuzzy Logic Controller (FLC), and Artificial Neural Network (ANN), have been proposed for LFC [4], [5], [6], [7], [8] and [9]. Design of a controller for AGC can be divided into two groups. In the 1st group the controller gains are tuned by a suitable optimization algorithm. In the 2nd group researchers have adopted self-tuning techniques with the help of neural network and fuzzy logic. Fuzzy logic controllers have been successfully used for analysis and control of non-linear system in the past decades. Yesil et al. [10] have used a self-tuning fuzzy PID type controller for load frequency control of a two-area interconnected system. Khuntia and Panda [11] have used ANFIS approach for AGC of a three area system. Ghosal [8] have used PSO optimization technique to optimize the PID controller gain for a fuzzy based LFC. These methods provide good performances but the transient responses are oscillatory in nature. Fuzzy logic based PID controller can be successfully used for all non-linear system but there is no specific mathematical formulation to decide the proper choice of fuzzy parameters (such as inputs, scaling factors, membership functions, and rule base). Normally these parameters are selected by using certain empirical rules and therefore may not be the optimal parameters. Improper selection of input–output scaling factor may affect the performance of FLC to a greater extent.

To get an accurate insight into the AGC problem, it is necessary to include the important physical constraints in the system model. The major physical constraints that affect the power system performance are Generation Rate Constraint (GRC) and time delay. The Flexible AC Transmission System (FACTS) controllers [12] play a crucial role to enhance power system stability in addition to control the power flow in an interconnected power system. Several studies have explored the potential of using FACTS devices for better power system control since it provides more flexibility. A Superconducting Magnetic Energy Storage (SMES) is capable of controlling both active and reactive power simultaneously. SMES unit with small storage capacity can be essential not only as a fast energy compensation device for power consumptions of large loads, but also as a stabilizer of frequency oscillations [13]. TCSC is one of the FACTS controller which is enhanced the power system dynamics, power transfer capability of transmission lines and dynamic stability [14].

It obvious from the literature survey that the performance of the power system not only depends on the controller structure but also depends on the artificial optimization technique. Hence, proposing and implementing new high performance heuristic optimization algorithms to real world problems are always welcome. Differential Evolution (DE) is a population-based direct search algorithm for global optimization capable of handling non-differentiable, non-linear and multi-modal objective functions, with few, easily chosen, control parameters [15] and [16]. However, the success of DE in solving a specific problem crucially depends on appropriately choosing trial vector generation strategies and their associated control parameter values namely the step size F, crossover probability CR, number of population NP and generations G [17].

In view of the above, a Differential Evolution (DE) optimized fuzzy PID controller is proposed for Load Frequency Control (LFC) of multi-area multi-units thermal–thermal power system with the consideration of reheat turbine, Generation Rate Constraint (GRC) and time delay. The superiority of the proposed approach is shown by comparing the results with GA for the same power system. Further, TCSC is employed in series with the tie-line in coordination with SMES to improve the dynamic performance of the power system. Finally, sensitivity analysis is carried out by varying the loading condition and system parameters.

MATERIALS AND METHODS

System under study

The system under investigation consists of two area interconnected thermal power system as shown in Fig. 1. Area 1 comprises two reheat thermal power units. Area 2 comprises two non-reheat thermal units. In Fig. 1, B_1 and B_2 are the frequency bias parameters; ACE_1 and ACE_2 are area control errors; R_1, R_2 and R_3, R_4 are the governor speed regulation parameters in pu Hz for area 1 and area 2 respectively; T_{G1}, T_{G2} and T_{G3}, T_{G4} are the speed governor time constants in sec for area 1 and area 2 respectively; T_{T1}, T_{T2} and T_{T3}, T_{T4} are the turbine time constant in sec for area 1 and area 2 respectively; ΔP_{D1} and ΔP_{D2} are the load demand changes; ΔP_{Tie} is the incremental change in tie line power (p.u); K_{Ps1} and K_{Ps2} are the power system gains; T_{Ps1} and T_{Ps2} are the power system time constant in sec; T_{12} is the synchronizing coefficient and ΔF_1 and ΔF_2 are the system frequency deviations in Hz. To get an accurate insight into the AGC problem, it is essential to include the important inherent requirement and the basic physical constraints and include them model. The important constraints that affect the power system performance are Generation Rate Constraint (GRC), and Time delay. In view of the above, the effect of GRC and Time delay are included to a power system model. Time delays can degrade a system's performance and even cause system instability. In a power system having steam plants, power generation can change only at a specified maximum rate. In thermal power plants, power generation can change only at a specified maximum/minimum rate known as Generation Rate Constraint (GRC). In the present study, a GRC of 3%/min for reheat and 10%/ min for non-reheat thermal units are considered [18] and [19]. Also in the present study, a time delay of 50 ms is considered [20]. The relevant parameters are given in Appendix A.

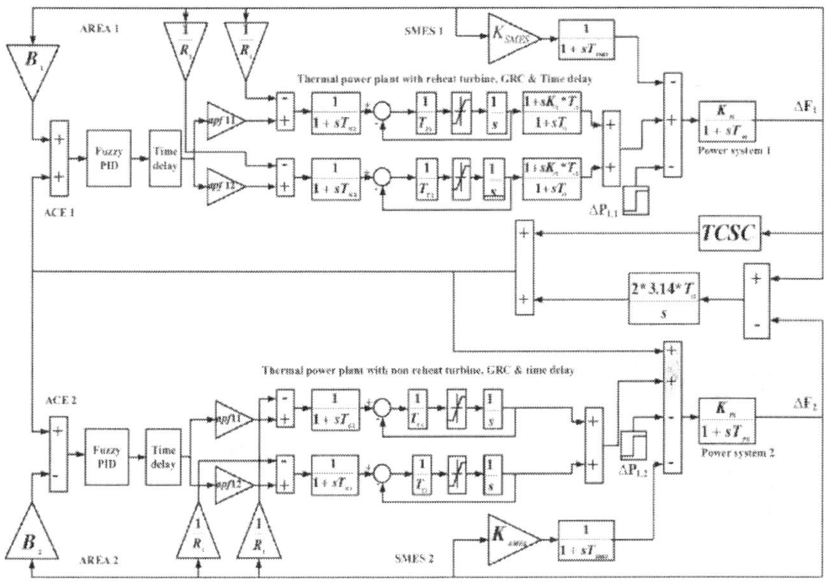

Figure 1. MATLAB/SIMULINK model of multi-area multi-units thermal system.

Control structure and objective function

To control the frequency, fuzzy PID controllers are provided in each area. The structure of fuzzy PID controller is shown in Fig. 2[21] and [22].

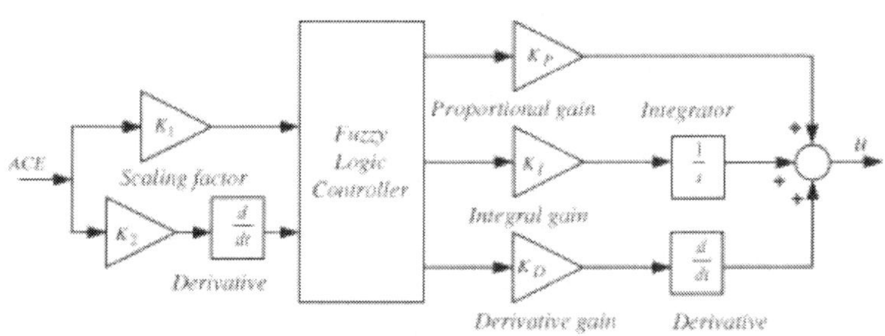

Figure 2. Structure of proposed fuzzy PID controller.

The error inputs to the controllers are the respective area control errors (ACE) given by:

$$e_1(t)=ACE_1=B_1\Delta F_1+\Delta P_{Tie} \quad (1)$$

$$e_2(t)=ACE_2=B_2\Delta F_2-\Delta P_{Tie} \quad (2)$$

Fuzzy controller uses error (e) and derivative of error $(\dot{e})$ as input signals. The outputs of the fuzzy controllers u_1 and u_2 are the control inputs of the power system i.e. the reference power settings ΔP_{ref1} and ΔP_{ref2}. The input scaling factors are the tuneable parameters K_1 and K_2. The proportional, integral and derivative gains of fuzzy controller are represented by K_P, K_I and K_D respectively. Triangular membership functions are used with five fuzzy linguistic variables such as NB (negative big), NS (negative small), Z (zero), PS (positive small) and PB (positive big) for both the inputs and the output. Membership functions for error, error derivative and FLC output are shown in Fig. 3. Mamdani fuzzy interface engine is selected for this work. The FLC output is determined by using center of gravity method of defuzzification. The two-dimensional rule base for error, error derivative and FLC output is shown in Table 1.

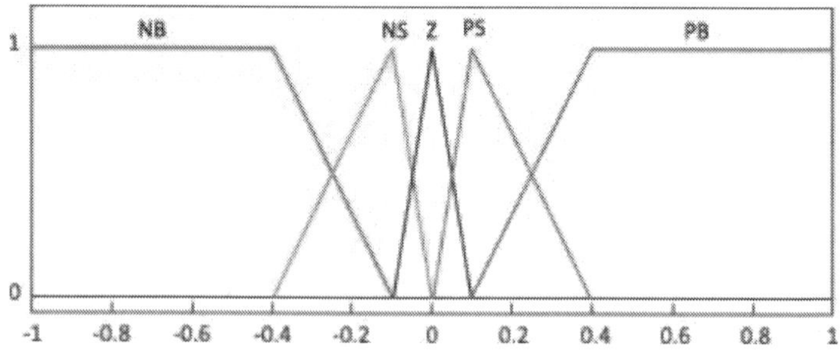

Figure 3. Membership functions for error, error derivative and FLC output.

Table 1. Rule base for error, derivative of error and FLC output.

	$\dot{e}$				
e	NB	NS	Z	PS	PB
NB	NB	NB	NS	NS	Z
NS	NB	NS	NS	Z	PS
Z	NS	NS	Z	PS	PS
PS	NS	Z	PS	PS	PB
PB	Z	PS	PS	PB	PB

In the design of modern heuristic optimization technique based controller, the objective function is first defined based on the desired specifications and constraints. Typical output specifications in the time domain are peak overshooting, rise time, settling time, and steady-state error. It has been reported in the literature that Integral of Time multiplied Absolute Error (ITAE) gives a better performance compared to other integral based performance criteria [23]. Therefore in this paper ITAE is used as objective function to optimize the scaling factors and proportional, integral and derivative gains of fuzzy PID controller. Expression for the ITAE objective function is depicted in Eq. (3).

$$J = ITAE = \int_0^{t_{sim}} (|\Delta F_1| + |\Delta F_2| + |\Delta P_{Tie}|) \cdot t \cdot dt \tag{3}$$

In the above equation, ΔF_1 and ΔF_2 are the system frequency deviations; ΔP_{Tie} is the incremental change in tie line power; t_{sim} is the time range of simulation.

Modeling of TCSC in AGC

It is well known that the reactance adjusting of Thyristor Controlled Series Compensator (TCSC) is a complex dynamic process. Effective design and accurate evaluation of the TCSC control strategy depends on the simulation accuracy of this process. Basically a TCSC consists of three components: capacitor banks, bypass inductor and bidirectional thyristors. The firing angles of the thyristors are controlled to adjust the TCSC reactance in accordance with a system control algorithm, normally in response to some system parameter variations. According to the variation in the thyristor firing angle, this process can be modeled as a fast switch between corresponding reactance offered to the power system. Both capacitive and inductive reactance compensation are possible by proper selection of capacitor and inductor values of the TCSC device. TCSC is considered as a variable reactance, the value of which is adjusted automatically to constrain the power flow across the branch to a specified value. The variable reactance X_{TCSC} represents the net equivalent reactance of the TCSC, when operating in either the inductive or the capacitive mode [14]. Fig. 4 shows the schematic diagram of a two area interconnected thermal-thermal power system with TCSC connected in series with the tie-line. For analysis, it is assumed that TCSC is connected near to the area 1. Resistance of the tie-line is neglected, since the effect on the dynamic performance is negligible. Further, the reactance to resistance ratio in a practically interconnected power system is quite high. The incremental tie-line power flow without TCSC is given by (4).

$$\Delta P_{Tie12}(s) = \frac{2\pi T_{12}^{0}}{s}\left[\Delta F_1(s) - \Delta F_2(s)\right] \tag{4}$$

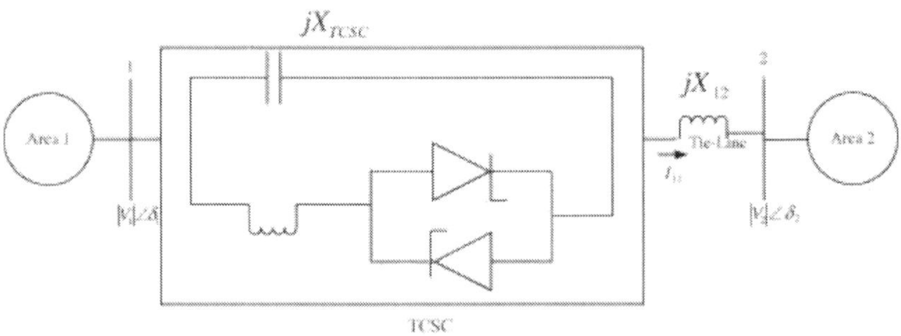

Figure 4. Two-area interconnected power system with TCSC.

In the above equation, ΔF_1 and ΔF_2 are the system frequency deviations; T_{12}^{0} is the synchronizing coefficient without TCSC. The line current flow from area-1 to area-2 can be written as, when TCSC is connected in series with the tie-line

$$I_{12} = \frac{|V_1|\angle(\delta_1) - |V_2|\angle(\delta_2)}{j(X_{12} - X_{TCSC})} \tag{5}$$

where X_{12} and X_{TCSC} are the tie-line and TCSC reactance respectively. It is clear from Fig. 4 that, the complex tie-line power as

$$P_{Tie12} - jQ_{Tie12} = V_1^{*} I_{12} = |V_1|\angle(-\delta_1)\left[\frac{|V_1|\angle(\delta_1) - |V_2|\angle(\delta_2)}{j(X_{12} - X_{TCSC})}\right] \tag{6}$$

Solving the above equation, the real part,

$$P_{Tie12} = \frac{|V_1||V_2|}{(X_{12} - X_{TCSC})} \sin(\delta_1 - \delta_2) \tag{7}$$

The tie-line power flow can be represented in terms of % compensation (k_c) as

$$P_{Tie12} = \frac{|V_1||V_2|}{X_{12}(1 - k_C)} \sin(\delta_1 - \delta_2) \tag{8}$$

where $k_c = \frac{X_{TCSC}}{X_{12}}$, percentage of compensation offered by the TCSC
In order to obtain the linear incremental model, Eq. (8) can be rewritten as

$$\Delta P_{Tie12} = \frac{|V_1||V_2|}{X_{12}(1 - k_C^0)^2} \sin(\delta_1^0 - \delta_2^0)\Delta k_C + \frac{|V_1||V_2|}{X_{12}(1 - k_C^0)} \cos(\delta_1^0 - \delta_2^0)(\Delta\delta_1 - \Delta\delta_2) \tag{9}$$

If $J_{12}^0 = \frac{|V_1||V_2|}{X_{12}} \sin(\delta_1^0 - \delta_2^0)$ and $T_{12}^0 = \frac{|V_1||V_2|}{X_{12}} \cos(\delta_1^0 - \delta_2^0)$, then Eq. (9) is expressed as

$$\Delta P_{Tie12} = \frac{J_{12}^0}{(1 - k_C^0)^2} \Delta k_C + \frac{T_{12}^0}{(1 - k_C^0)} (\Delta\delta_1 - \Delta\delta_2) \tag{10}$$

Since $\Delta\delta_1 = 2\pi \int \Delta F_1 dt$ and $\Delta\delta_2 = 2\pi \int \Delta F_2 dt$

Taking Laplace transforms of Eq. (10) and expressed as given by (11)

$$\Delta P_{Tie12}(s) = \frac{J_{12}^0}{(1 - k_C^0)^2} \Delta k_C(s) + \frac{2\pi T_{12}^0}{s(1 - k_C^0)} [\Delta F_1(s) - \Delta F_2(s)] \tag{11}$$

From Eq. (11), the tie-line power flow can be regulated by controlling $\Delta k_c(s)$. If the control input signal to TCSC damping controller is assumed to be ΔError(s) and the transfer function of the signal conditioning circuit is $k_c = \frac{K_{TCSC}}{1+sT_{TCSC}}$, The expression is given (12)

$$\Delta k_C(s) = \frac{K_{TCSC}}{1 + sT_{TCSC}} \Delta \mathrm{Error}(s) \tag{12}$$

where K_{TCSC} and T_{TCSC} is the gain and time constant of the TCSC controller respectively. As TCSC is kept near to area-1, frequency deviation ΔF_1 may be suitably used as the control signal ΔError(s), to the TCSC unit to control the percentage incremental change in the system compensation level. Therefore,

$$\Delta k_C(s) = \frac{K_{TCSC}}{1 + sT_{TCSC}} \Delta F_1(s) \tag{13}$$

$$\Delta P_{Tie12} = \frac{2\pi T_{12}^0}{s(1 - k_C^0)} [\Delta F_1(s) - \Delta F_2(s)] + \left[\frac{J_{12}^0}{(1 - k_C^0)^2}\right] \frac{K_{TCSC}}{1 + sT_{TCSC}} \Delta F_1(s) \tag{14}$$

Modeling of SMES in AGC

Superconducting Magnetic Energy Storage (SMES) is a device which can store the electrical power from the grid in the magnetic field of a coil. The magnetic field of coil is made of superconducting wire with near-zero loss of energy. SMESs can store and refurbish huge values of energy almost instantaneously. Therefore the power system can discharge high levels of power within a fraction of a cycle to avoid a rapid loss in the line power. The SMES is consisting of inductor-converter unit, dc superconducting inductor, AC/ DC converter and a step down transformer [24]. The stability of a SMES unit is superior to other power storage devices, because all parts of a SMES unit are static. Fig. 5 shows the schematic diagram of SMES unit in the power system [13]. During normal operation of the grid, the superconducting coil will be charged to a set value (normally less than the maximum charge) from the utility grid. After charged, the superconducting coil conducts current, which supports an electromagnetic field, with virtually no losses. The coil is kept at very low temperature by immersion in a bath of liquid helium.

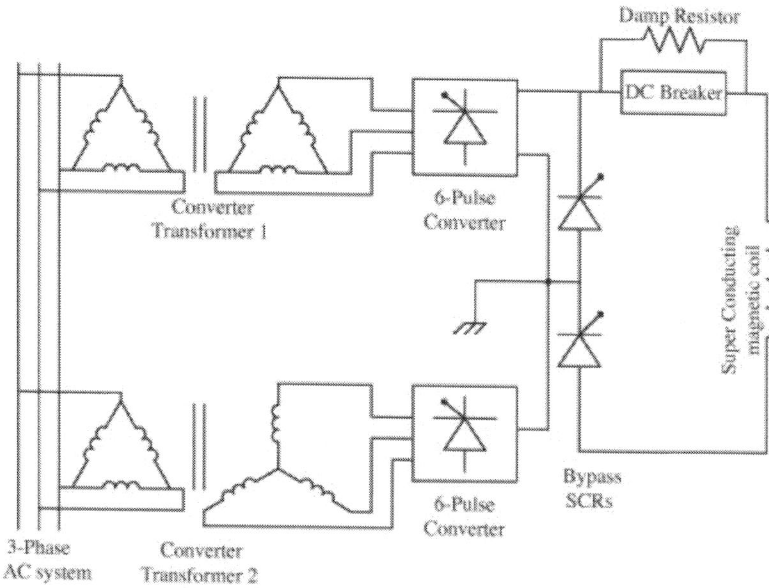

Figure 5. SMES circuit diagram.

In the present work two SMES units are established in area1 and area2 in order to stabilize frequency oscillations as shown in Fig. 1. The input signal of the SMES controller is p.u. frequency deviation (ΔF) and the output is the change in control vector [ΔP_{SMES}]. The controller gains K_{SMES} and the time constant T_{SMES} values are 0.12 and 0.03 s respectively [24].

Over view of differential evolution

Differential Evolution (DE) algorithm is a search heuristic algorithm introduced by Storn and Price [15]. It is a simple, efficient, reliable algorithm with easy coding. The main advantage of DE over Genetic Algorithm (GA) is that GA uses crossover operator for evolution while DE relies on mutation operation. The mutation operation in DE is based on the difference in randomly sampled pairs of solutions in the population. An optimization task consisting of D variables can be represented by a D-dimensional vector. A population of N_P solution vectors is randomly initialized within the parameter bounds at the beginning. The population is modified by applying mutation, crossover and selection operators. DE algorithm uses two generations; old generation and new generation of the same population size. Individuals of the current population become target vectors for the next generation. The mutation operation produces a mutant vector for each target vector, by adding the weighted difference between two randomly chosen vectors to a third vector. A trial vector is generated by the crossover operation by mixing the parameters of the mutant vector with those of the target vector. The trial vector substitutes the target vector in the next generation if it obtains a better fitness value than the target vector. The evolutionary operators are described below [25] and [26]:

Initialization of parameter

DE begins with a randomly initiated population of size N_P *of* D dimensional real-valued parameter vectors. Each parameter j lies within a range and the initial population should spread over this range as much as possible by uniformly randomizing individuals within the search space constrained by the prescribed lower bound X_j^L and upper bound X_j^U.

Mutation operation

For the mutation operation, a parent vector from the current generation is selected (known as target vector), a mutant vector is obtained by the differential mutation operation (known as donor vector) and finally an offspring is produced by combining the donor with the target vector (known as trial vector). Mathematically it can be expressed as:

$$V_{i,G+1} = X_{r1,G} + F.(X_{r2,G} - X_{r3,G}) \tag{15}$$

where $X_{i,G}$ is the given parameter vector, $X_{r1,G}$ $X_{r2,G}$ $X_{r3,G}$ are randomly selected vector with distinct indices i, $r1$, $r2$ and $r3$, $V_{i,G+1}$ is the donor vector and F is a constant from (0, 2)

Crossover operation

After generating the donor vector through mutation the crossover operation is employed to enhance the potential diversity of the population. For crossover operation three parents are selected and the child is obtained by means of perturbation of one of them. In crossover operation a trial vector $U_{i,G+1}$ is obtained from target vector ($X_{i,G}$) and donor vector ($V_{i,G}$). The donor vector enters the trial vector with probability CR given by:

$$U_{j,i,G+1} = \begin{cases} V_{j,i,G+1} & if\ rand_{j,i} \leqslant CR \quad or \quad j = I_{rand} \\ X_{j,i,G+1} & if\ rand_{j,i} > CR \quad or \quad j \neq I_{rand} \end{cases}$$

(16)

With $rand_{j,i} \sim U(0, 1)$, I_{rand} is a random integer from (1, 2, …, D) where D is the solution's dimension i.e. number of control variables. I_{rand} ensures that $v_{i,G+1} \neq x_{i,G}$.

Selection operation

To keep the population size constant over subsequent generations, selection operation is performed. In this operation the target vector $X_{i,G}$ is compared with the trial vector $v_{i,G+1}$ and the one with the better fitness value is admitted to the next generation. The selection operation in DE can be represented by:

$$X_{i,G+1} = \begin{cases} U_{i,G+1} & if\ f(U_{i,G+1}) < f(X_{i,G}) \\ X_{i,G} & otherwise. \end{cases}$$

(17)

where $i \in [1, NP]$.

RESULTS AND DISCUSSIONS

Implementation of DE

The effectiveness, efficiency, and robustness of the DE algorithm are sensitive to the settings of the control parameters. The control parameters in DE are step size function also called scaling factor (F), crossover probability (CR), the number of population (N_P), initialization, termination and evaluation function. F controls the amount of perturbation in the mutation process and generally lies in the range (0, 1). Crossover probability (CR) constants are generally chosen from the interval (0.5, 1). Several strategies can be employed in DE optimization algorithm. The strategy in a DE algorithm is denoted by DE/x/y/z, where x represents the mutant vectors, y represents the number of difference vectors used in the mutation process and z represents the crossover scheme used in the crossover operation. The suggested choice of control parameters is [25] population size of $N_P = 50$ ($N_P = 5D$ where D = dimensionality of the problem), step size $F = 0.8$ and crossover probability of $CR = 0.8$ and these values are selected in the present paper. The strategy employed is as follows: DE/best/1/exp. Optimization is terminated by the pre-specified number of generations which is set to 100. The flow chart of the DE algorithm employed in the present study is given in Fig. 6. The model of the system under study shown in Fig. 1 is developed in MATLAB/SIMULINK environment and DE program is written (in .mfile). Initially, fuzzy PID controllers without TCSC and SMES units are considered for each area. Scaling factors and PID controller gains are chosen in the range [0 −2] and [−2 2] respectively. The developed model is simulated in a separate program (by .mfile using initial population/controller parameters) considering a 1% step load change in area 1. The objective function (ITAE) value for each individual is calculated in the SIMULINK model file and transferred to .mfile through workspace. These objective function values are used to assess the populations. The population is then modified by applying mutation, crossover and selection operators in the main DE program as given in Flow chart (Fig. 6). Simulations were conducted on an Intel, core i-3core cpu, of 2.4 GHz and 4 GB RAM computer in the MATLAB 7.10.0.499 (R2010a) environment. The optimization was repeated 50 times and the best final solution among the 50 runs is chosen as proposed controller parameters. The best final solutions obtained in the 50 runs are shown in Table 2.

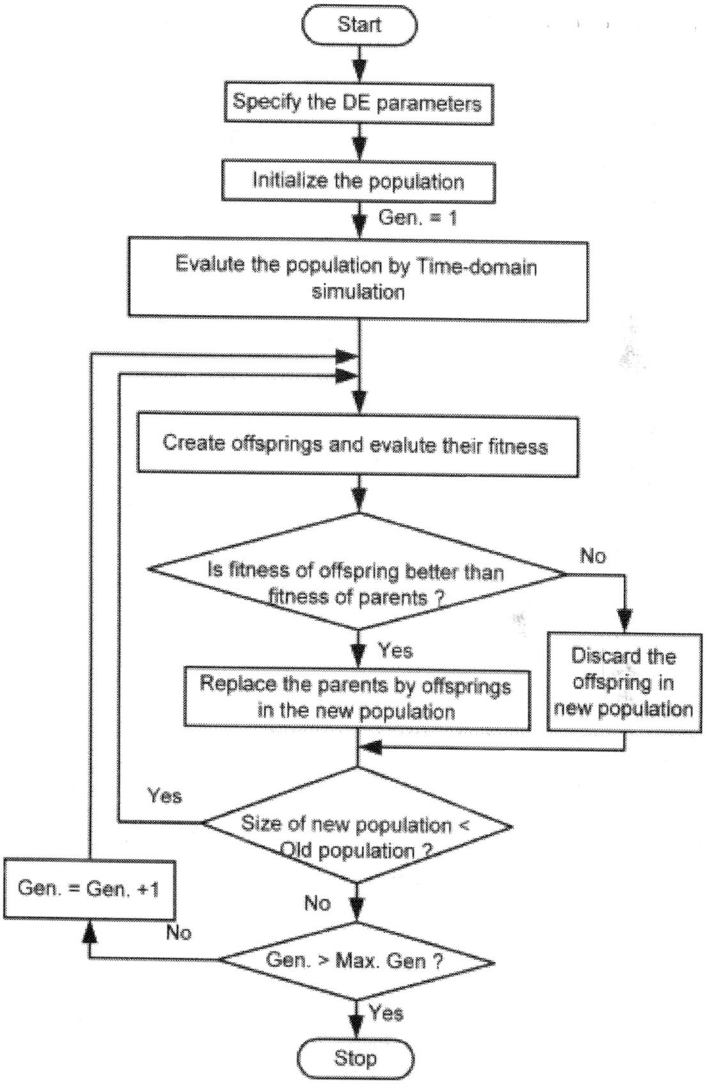

Figure 6. Flow chart of proposed DE optimization approach.

Table 2. Tuned fuzzy PID controller parameters.

	Genetic Algorithm (GA)	Differential Evolution (DE)		
Optimum controller gains	Without TCSC and SMES	Without TCSC and SMES	With TCSC	With TCSC and SMES
K_1	1.5553	1.8589	0.1943	0.1132
K_2	1.792	1.9092	1.5638	1.682
K_{P1}	−0.0805	0.5481	1.672	1.5621
K_{I1}	1.346	1.9461	1.7881	1.5075
K_{D1}	1.3483	0.517	−0.6707	−0.3560
K_3	1.9128	0.3338	0.5414	0.1932
K_4	0.2626	0.3607	0.2572	0.3662
K_{P2}	0.2813	− 0.2837	−0.9896	−1.7376
K_{I2}	1.1323	−0.8919	−0.2052	1.3894
K_{D2}	−0.0970	1.3173	0.9608	−1.0557

Analysis of results

The objective function (ITAE) value given by Eq. (3) is determined by simulating the developed model by applying a 1% step increase in load in area 1. The corresponding performance index in terms of ITAE value, settling times (2%) and peak overshoots in frequency and tie line power deviations is shown in Table 3. For comparison, the corresponding values of GA optimized fuzzy PID controllers are also shown in Table 3. For the implementation of GA, normal geometric selection, arithmetic crossover and non-uniform mutation are employed in the present study. A population size of 50 and maximum generation of 100 is employed in the present paper. A detailed description about GA parameters employed in the present paper can be found in reference [9]. It should be noted here that, GA values correspond to same power system, controller structure (fuzzy PID) and objective function employed (ITAE) for proper comparison of techniques. It is evident from Table 3 that DE outperform GA as minimum ITAE value is obtained with DE (ITAE = 1.2250) compared to GA (ITAE = 2.1429). The dynamic performance of the system is shown in Figure 7, Figure 8 and Figure 9 for 1% step increase in load in area 1. It is clear from Figure 7, Figure 8 and Figure 9 that better dynamic performance is obtained by DE optimized fuzzy PID controller compared to GA optimized fuzzy PID controller. Hence it can be concluded that DE outperform GA technique.

Table 3. Comparative performance of error and settling time.

Parameters	ITAE	Settling time (2% band), T_s (s)			Peak over shoot ($\times 10^{-3}$)		
		ΔF_1	ΔF_2	ΔP_{Tie}	ΔF_1	ΔF_2	ΔP_{Tie}
GA	2.1429	37.05	34.85	34.06	6.413	1.568	0.839
DE	1.225	33.26	35.03	30.43	3.974	1.528	0.756
DE: with TCSC	0.8178	21.84	22.25	26.25	3.768	8.01	0.762
DE: both TCSC and SMES	**0.6672**	**21.13**	**21.57**	**16.85**	**2.5113**	**5.44**	**0.29**

The bold values are indicates the best results.

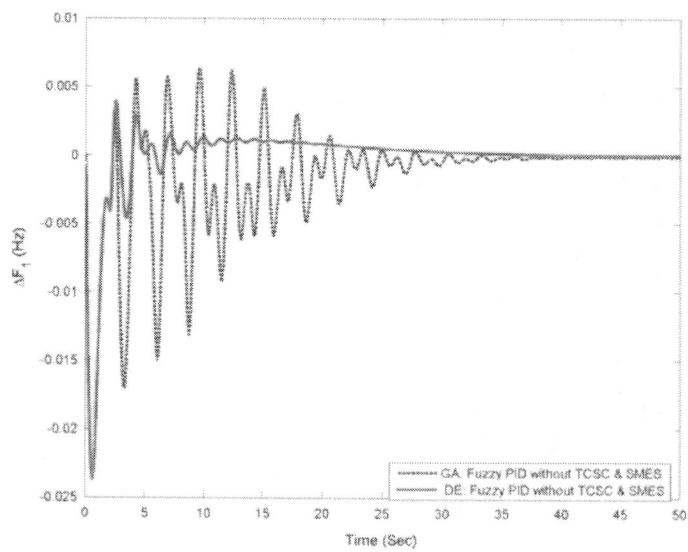

Figure 7. Change in frequency of area-1 for 1% load change in area-1 without TCSC and SMES units.

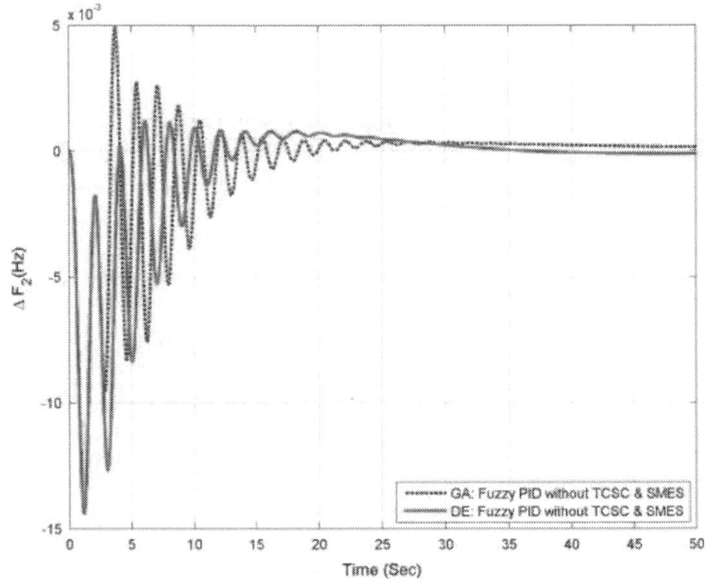

Figure 8. Change in frequency of area-2 for 1% load change in area-1 without TCSC and SMES units.

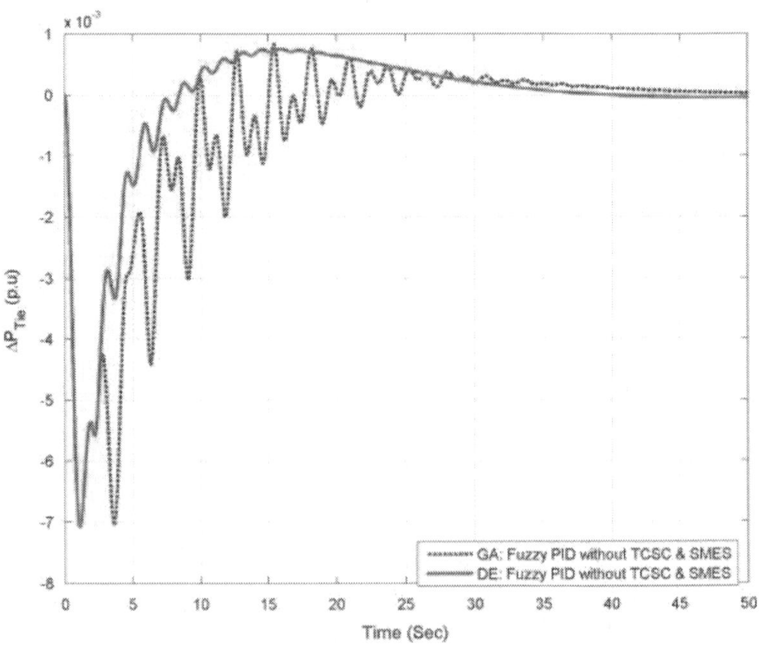

Figure 9. Change in tie line power for 1% load change in area-1 without TCSC and SMES units.

In the next step, the TCSC is incorporated separately in the tie-line to analysis its effect on the power system performance. Subsequently SMES units are installed in both areas and coordinated with TCSC to study their effect on system performance. The results of fuzzy PID controller with TCSC employing differential evolution algorithm over 50 independent runs are shown in Table 2. It is clear from Table 3 that by employing the TCSC along with fuzzy PID controller, the objective function (ITAE) value is decreased to 0.8178 (i.e. 33.24% improvement). In addition better results are observed in terms of settling time and peak overshoot values with the TCSC fuzzy PID compared to without TCSC. It is also seen that with coordinated application of TCSC and SMES units, the ITAE value is further reduced to **0.6672**. It can be seen from Table 3 that with TCSC and SMES, the settling times of ΔF_1, ΔF_2 and ΔP_{Tie} are improved compared to others for the same investigated system with similar objective function (ITAE).

To study the dynamic performance of the system a step increase in demand of 1% is applied at $t = 0$ s in area-1 and the system dynamic responses are shown in Figure 10,Figure 11 and Figure 12. Critical analysis of the dynamic responses clearly reveals that significant system performance improvement in terms of minimum undershoot and overshoot in frequency oscillations as well as tie-line power exchange is observed with coordinated application of TCSC and SMES units.

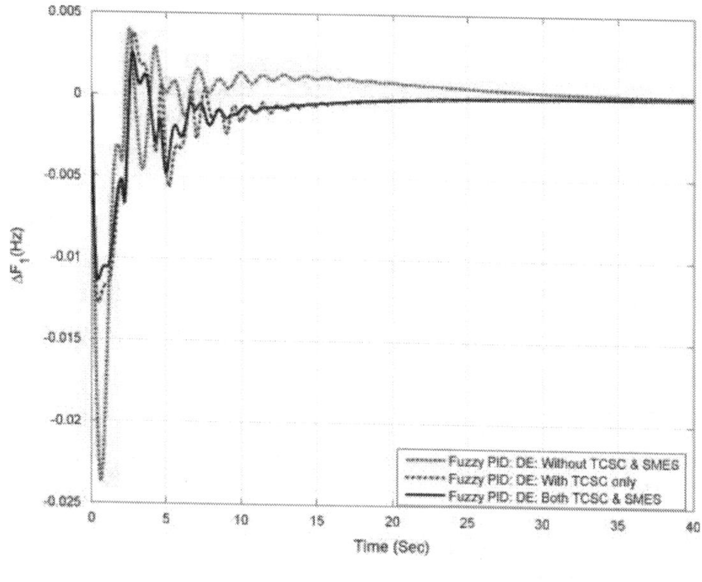

Figure 10. Change in frequency of area-1 for 1% load change in area-1.

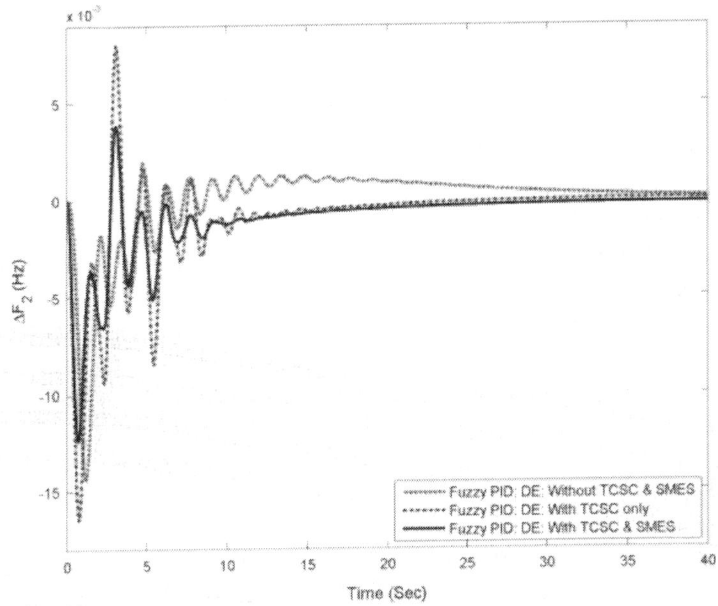

Figure 11. Change in frequency of area-2 for 1% load change in area-1.

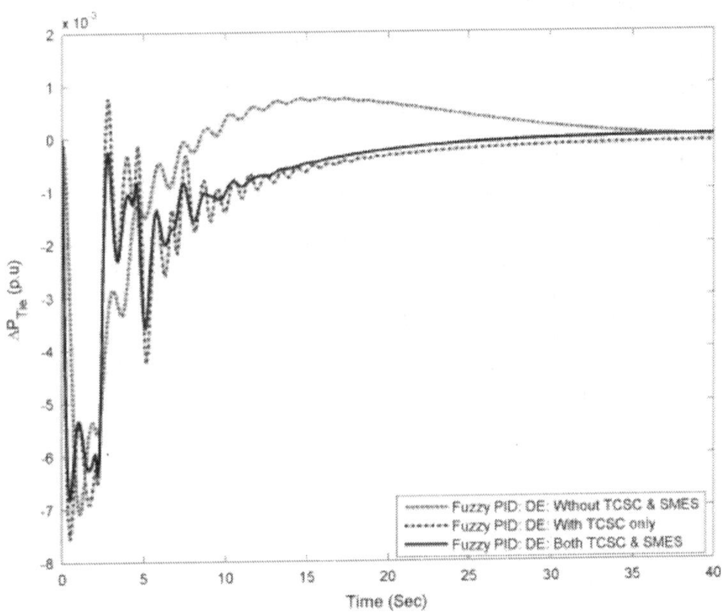

Figure 12. Tie-line power deviation for 1% load change in area-1.

Sensitivity analysis

Sensitivity analysis is carried out to study the robustness the system to wide changes in the operating conditions and system parameters [5], [27] and [28]. In this section robustness of the power system is checked by varying the loading conditions and system parameters from their nominal values (given in Appendix A) in the range of +25% to −25% without changing the optimum values of fuzzy PID controller gains. The change in operating load condition affects the power system parameters K_P and T_P. The power system parameters are calculated for different loading conditions as given in Appendix A. The system with TCSC and SMES units are considered in all the cases due to their superior performance. The various performance indexes (settling time, peak overshoot and ITAE) under normal and parameter variation cases for the system are given in Table 4. It can be observed from Table 4 that settling time, peak overshoot and ITAE values vary within acceptable ranges and are nearby equal to the respective values obtained with nominal system parameter. It is also evident from Table 5 and Table 6 that the eigenvalues lie in the left half of s-plane for all the cases thus maintain the stability. Hence, it can be concluded that the proposed controllers are robust and perform satisfactorily when system parameters changes in the range ±25%. The dynamic performance of the system with the varied conditions of loading, T_G, T_T, B and R is shown in Figure 13, Figure 14, Figure 15, Figure 16, Figure 17, Figure 18 and Figure 19. It can be observed from Figure 13, Figure 14, Figure 15, Figure 16, Figure 17, Figure 18 and Figure 19 that the effect of the variation in loading condition and system parameters on the system performance is negligible. Hence the optimum values of controller parameters obtained at the nominal loading with nominal parameters, need not be reset for wide changes in the system loading or system parameters.

Table 4.Sensitivity analysis.

Parameter variation	% Change	Performance index with TCSC and SMES						ITAE
		Settling time, T_s (s)			Peak over shoot × 10^{-3}			
		ΔF_1	ΔF_2	ΔP_{Tie}	ΔF_1	ΔF_2	ΔP_{Tie}	
Nominal	**0**	**21.13**	**21.57**	**16.85**	**2.5113**	**5.44**	**0.29**	**0.6672**
Loading condition	25	21.83	22.22	17.15	2.489	5.364	0.266	0.665
	−25	20.47	20.88	16.4	2.533	5.517	0.313	0.6825
T_G	25	23.03	23.45	17.59	2.835	5.895	0.488	0.7144
	−25	19.31	19.76	15.87	2.147	4.946	0.171	0.7351
T_t	25	19.44	19.87	15.94	2.564	5.48	0.298	0.7203
	−25	23.29	23.68	17.68	2.395	5.291	0.233	0.7256
B	25	20.63	21.1	17.37	2.493	5.53	0.299	0.6492
	−25	21.65	22.04	16.14	2.531	5.354	0.28	0.6912
R	25	21.48	21.89	17.11	2.494	5.485	0.194	0.7069
	−25	20.68	21.13	16.32	2.514	5.362	0.422	0.6103

The bold values are indicates the best results.

Table 5. System eigen values under parameter variation in loading, TG and TT with TCSC and SMES units.

Loading condition		T_G		T_T	
25%	−25%	25%	−25%	25%	−25%
−48.2086	−48.2096	−48.2093	−48.2086	−48.2093	−48.2088
−31.9578	−31.9592	−31.9596	−31.9561	−31.9593	−31.9572
−33.0234	−33.0237	−33.0236	−33.0236	−33.0236	−33.0236
−13.4364	−13.4354	−13.4629	−16.9560	−13.4497	−13.4189 ± 0.0790i
−13.2648	−13.2628	−12.9373	−13.4694	−13.1737	−1.4553 ± 4.7764i
−1.3825 ± 4.7594i	−1.3708 ± 4.7614i	−10.3946	−12.8478	−1.3341 ± 4.7430i	−1.2922 ± 3.0029i
−1.3825 ± 2.8736i	−1.2456 ± 2.8684i	−1.3570 ± 4.7719i	−1.3943 ± 4.7438i	−1.2419 ± 2.7746i	−2.7612
−1.3360	−1.3376	−1.2261 ± 2.8569i	−1.2809 ± 2.8822i	−2.1386	−1.3774
−0.1269	−0.1271	−1.3324	−1.3411	−1.2845	−0.1269
−0.0107	−0.0107	−0.1270	−0.1270	−0.1271	−0.0107
−0.0047	−0.0047	−0.0107	−0.0107	−0.0107	−0.0047
−0.1000	−0.1000	−0.0047	−0.0047	−0.0047	−2.3810
−12.4999	−12.5000	−12.5000	−12.5000	−2.3810	−0.1000
−12.5000	−12.5000	−0.1000	−0.1000	−0.1000	−12.5000
−2.3809	−2.3810	−2.3810	−2.3810	−12.5000	−12.5000
−2.3809	−2.3810	−2.3810	−2.3810	−12.5000	

Table 6. System eigen values under parameter variation in *B* and *R* with TCSC and SMES units.

B		*R*	
25%	−25%	25%	−25%
−48.2091	−48.2091	−48.2093	−48.2088
−31.9620	−31.9556	−31.9592	−31.9574
−32.9426	−33.1037	−33.0236	−33.0236
−13.4426	−13.4293	−13.4482	−13.4023 ± 0.0620i
−13.2631	−13.2646	−13.1869	−1.4277 ± 4.8201i
−1.3964 ± 4.7793i	−1.3573 ± 4.7410i	−1.3443 ± 4.7293i	−1.1697 ± 2.9829i
−1.2685 ± 2.8656i	−1.2363 ± 2.8767i	−1.3022 ± 2.7924i	−1.2936
−1.3362	−1.3374	−1.3679	−0.1305
−0.1270	−0.1270	−0.1247	−0.0109
−0.0051	−0.0104	−0.0106	−0.0043
−0.0111	−0.0043	−0.0050	−0.1000
−0.1000	−0.1000	−0.1000	−12.5000
−12.5000	−12.5000	−12.5000	−12.5000
−12.5000	−12.5000	−12.5000	−2.3810
−2.3810	−2.3810	−2.3810	−2.3810
−2.3810	−2.3810	−2.3810	

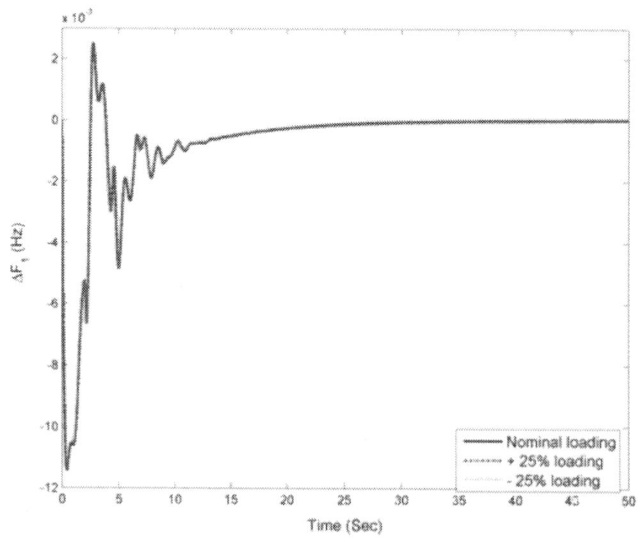

Figure 13. Change in frequency of area-1 for 1% load change in area-1 with variation in loading.

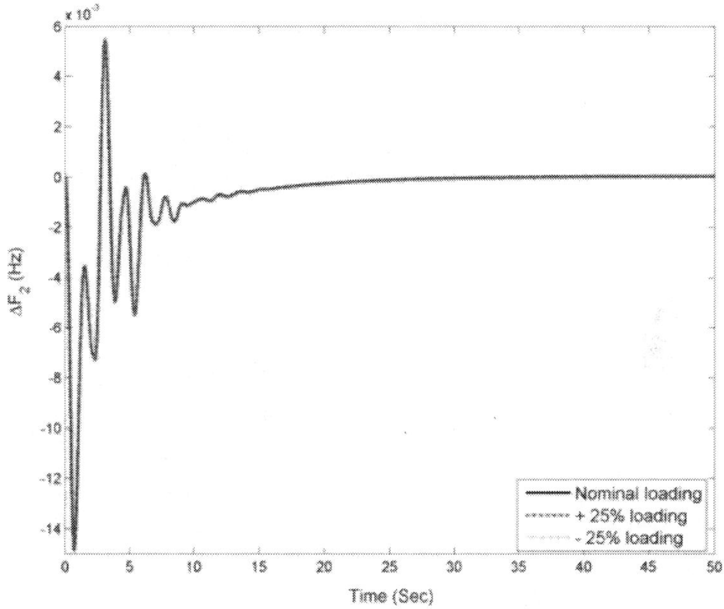

Figure 14. Change in frequency of area-2 for 1% load change in area-1 with variation in loading.

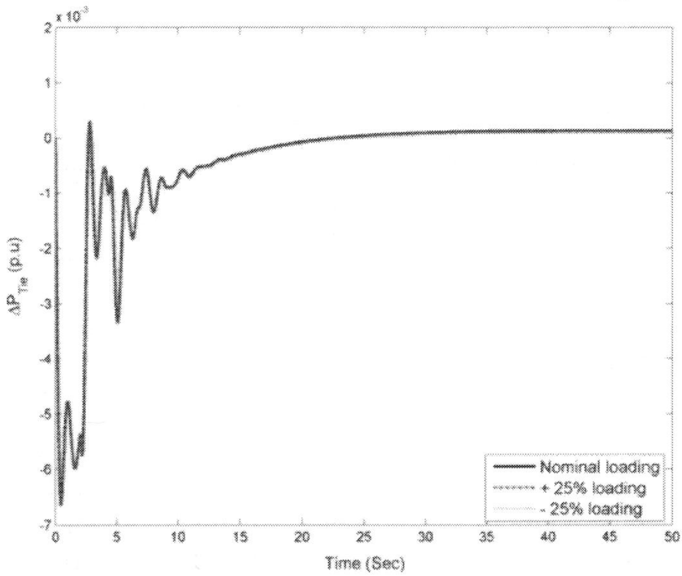

Figure 15. Tie-line power deviation for 1% load change in area-1 with variation in loading.

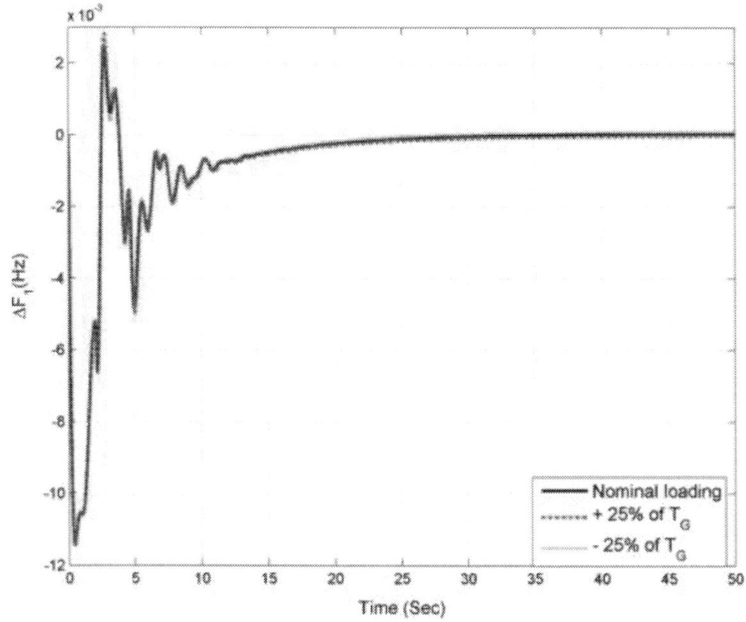

Figure 16. Change in frequency of area-1 for 1% load change in area-1 with variation in T_G.

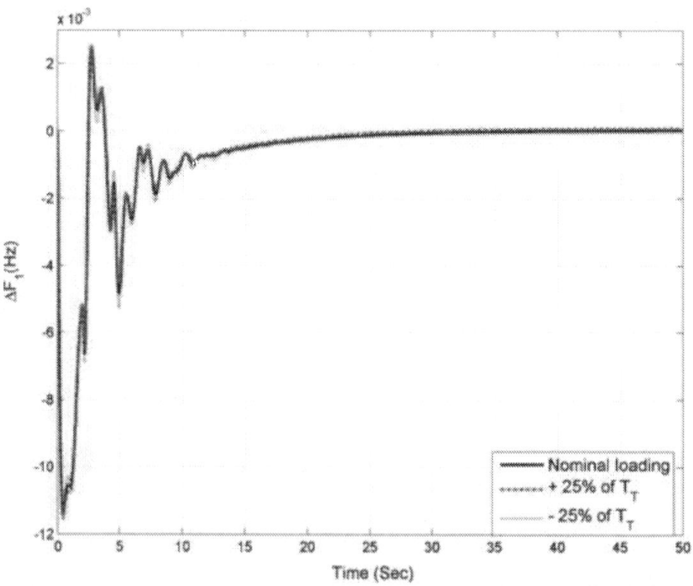

Figure 17. Change in frequency of area-1 for 1% load change in area-1 with variation in T_T.

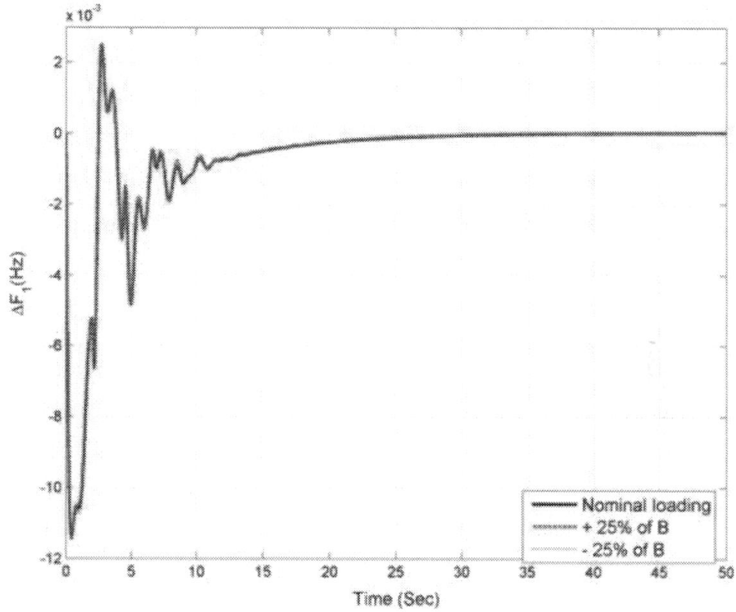

Figure 18. Change in frequency of area-1 for 1% load change in area-1 with variation in *B*.

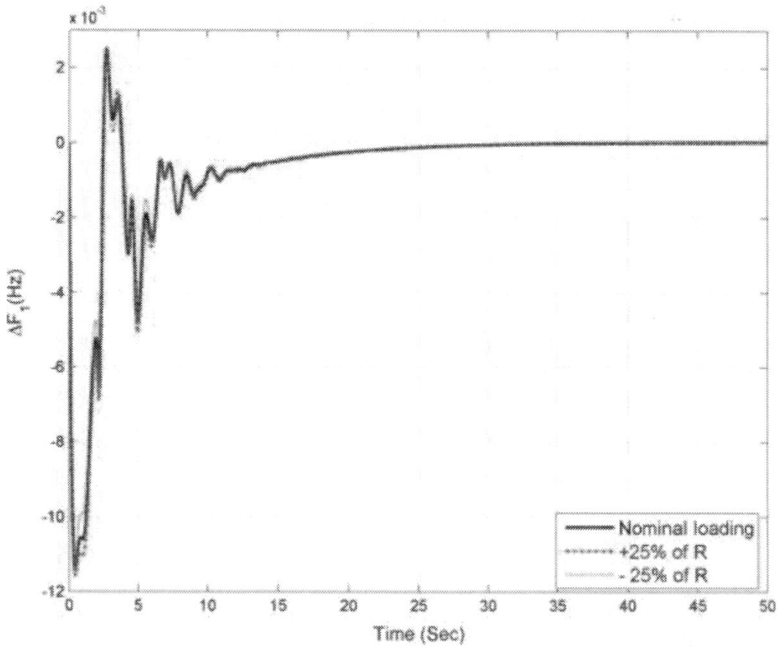

Figure 19. Change in frequency of area-1 for 1% load change in area-1 with variation in *R*.

CONCLUSION

In this paper, a Differential Evolution (DE) algorithm optimized fuzzy PID controller has been proposed for Automatic Generation Control (AGC) of multi-area multi-units power systems. Initially a multi-area multi-units power system with the considerations of physical constraints such as GRC and time delays is considered and the superiority of DE over GA is demonstrated. A linear incremental model for a TCSC has also been developed which is suitable for AGC applications. Further, TCSC and SMES units are added in the system model in order to improve the system performance. It is observed that when the TCSC unit is placed with the tie-line, dynamic performance of system is improved. Then the impact of SMES units in the AGC along with TCSC is studied. From the simulation results, it is observed that significant improvements of dynamic responses are obtained with coordinated application of TCSC and SMES units. Finally, sensitivity analysis is carried out to show the robustness of the controller by varying the loading conditions and system parameters in the range of +25% to −25% from their nominal values. For systems under study, it is revealed that the parameters of the proposed DE optimized fuzzy PID controllers need not be reset even if the system is subjected to wide variation in loading conditions and system parameters.

APPENDIX A

Nominal parameters of the system investigated are:

(i) **Multi-area multi-units system**

$B_1, B_2 = 0.42249$ p.u. MW/Hz; $R_1 = R_2 = R_3 = R_4 = 2.4$ Hz/p.u.; $T_{G1} = T_{G2} = T_{G3} = T_{G4} = 0.08$ s; $T_{T1} = T_{T2} = T_{T3} = T_{T4} = 0.3$ s; $K_P = 120$ Hz/p.u.; TP = 20 s;$K_{R1} = K_{R2} = 10$; $T_{R1} = T_{R2} = 10$ s

(ii) **TCSC data**

$T_{12} = 0.0866$; $\delta_0 = 300$; $X_t = 10$ p.u.; $K_{TCSC} = 2.0$;
$T_{TCSC} = 0.02$ s

(iii) **SMES data**

$K_{SMES} = 0.12$; $T_{SMES} = 0.03$ s

REFERENCES

1. Kundur P. Power system stability and control. New York: McGraw-Hill; 1994.
2. Elgerd OI. Electric energy systems theory – an introduction. 2nd ed. Tata McGraw Hill; 2007.
3. Kothari DP, Nagrath IJ. Modern power system analysis. 4th ed. New Delhi: Tata McGraw-Hill; 2011.
4. Saikia LC, Nanda J, Mishra S. Performance comparison of several classical controllers in AGC for multi-area interconnected thermal system. Int J Electr Power Energy Syst 2011;33:394–401.
5. Parmar KPS, Majhi S, Kothari DP. Load frequency control of a realistic power system with multi-source power generation. Int J Electr Power Energy Syst 2012;42:426–33.
6. Saikia LC, Mishra S, Sinha N, Nanda J. Automatic generation control of a multi area ydrothermal system using reinforced learning neural network controller. Int J Electr Power Energy Syst 2011;33(4):1101–8.
7. Ibraheem KP, Kothari DP. Recent philosophies of automatic generation control strategies in power systems. IEEE Trans Power Syst 2005;20(1):346–57.
8. Ghosal SP. Optimization of PID gains by particle swarm optimization in fuzzy based automatic generation control. Electr Power Syst Res 2004;72(3):203–12.
9. Golpira H, Bevrani H, Golpira H. Application of GA optimization for automatic generation control design in an interconnected power system. Energy Convers Manage 2011;52:2247–55.
10. Yesil E, Guzelkaya M, Eksin I. Self tuning fuzzy PID type load and frequency controller. Energy Convers Manage 2004;45(3):377–90.

11. Khuntia SR, Panda S. Simulation study for automatic generation control of a multi-area power system by ANFIS approach. Appl Soft Comput 2012;12(1):333–41.
12. Hingorani NG, Gyugyi L. Understanding FACTS-concepts and technology of flexible AC transmission systems. Standard Publishers, IEEE Press; 2000.
13. Praghnesh B, Ghoshal SP, Ranjit R. Load frequency stabilization by coordinated control of thyristor controlled phase shifters and superconducting magnetic energy storage for three types of interconnected two-area power systems. Int J Electr Power Energy Syst 2010;32:1111–24.
14. Mathur RM, Varma RK. Thyristor-based FACTS controllers for electrical transmission systems. IEEE Press, John Wiley & Sons, inc. publication; 2002.
15. Stron R, Price K. Differential evolution – a simple and efficient adaptive scheme for global optimization over continuous spaces. J Global Optim 1995;11:341–59.
16. Das S, Suganthan PN. Differential evolution: a survey of the state-of-the-art. IEEE Trans Evol Comput 2011;15:4–31.
17. Brest J, Greiner S, Boskovic B, Mernik M, Zumer V. Selfadapting control parameters in differential evolution: a comparative study on numerical benchmark problems. IEEE Trans Evol Comput 2005;10:646–57.
18. Cheres E. The application of generation rate constraint in modeling of a thermal power system. Electr Power Comp Syst 2001;29(2):83–7.
19. Ignacio E, Fernandez-Bernal F, Rouco L, Elosia P, Saiz-Chicharro A. Modeling of thermal generating units for automatic generation control purposes. IEEE Trans Control Syst Technol 2004;12(1):205–10.
20. Panda S. Differential evolution algorithm for SSSC-based damping controller design considering time delay. J. Franklin Inst 2011;348(8):1903–26.
21. Mudi KR, Pal RN. A robust self-tuning scheme for PI-and PDtype fuzzy controllers. IEEE Trans Fuzzy Syst 1999;7(1):2–16.
22. Woo WZ, Chung YH, Lin JJ. A PID type fuzzy controller with self tuning scaling factors. Fuzzy Sets Syst 2000;115(2):321–6.
23. Shabani H, Vahidi B, Ebrahimpour M. A robust PID controller based on imperialist competitive algorithm for load–frequency control of power systems. ISA Trans. 2012;52:88–95.
24. Sudha KR, Vijaya SR. Load frequency control of an interconnected reheat thermal system using type-2 fuzzy system including SMES units. Int J Electr Power Energy Syst 2012;43:1383–92.
25. Janez B, Saso G, Borko B, Marjan M, Viljem Z. Self-adapting control parameters in differential evolution: a comparative study on numerical benchmark problems. IEEE Trans Evol Comput 2005;10:646–57.

26. Qin AK, Huang VL, Suganthan PN. Differential evolution algorithm with strategy adaptation for global numerical optimization. IEEE Trans Evol Comput 2009;13:398–417.
27. Sahu RK, Panda S, Rout UK. DE optimized parallel 2-DOF PID controller for load frequency control of power system with governor dead-band nonlinearity. Int J Electr Power Energy Syst 2013;49:19–33.
28. Rout UK, Sahu RK, Panda S. Design and analysis of differential evolution algorithm based automatic generation control for interconnected power system. Ain Shams Eng J 2013;4(3):409–21.

CITATION

Saroj Padhan, Rabindra Kumar Sahu, Sidhartha Panda, Automatic generation control with thyristor controlled series compensator including superconducting magnetic energy storage units, Ain Shams Engineering Journal, Volume 5, Issue 3, September 2014, Pages 759-774, ISSN 2090-4479, http://dx.doi.org/10.1016/j.asej.2014.03.011.

Index

A

Annealing, 245, 246, 247, 248, 249, 250, 251, 252, 253, 254, 255, 256, 257, 258, 259, 262
Atomic force microscopy (AFM), 227, 228
Atomic layer deposition (ALD), 3
Atomic Layer Deposition (ALD), 247

B

Band-to-band transition, 125, 148
Bias voltage, 210, 222

C

Cap layer, 245
Capping layer, 227, 228, 230, 231, 232, 235, 241
Channel material, 246
Chemical vapor deposition (CVD), 3
Clipped signals with relaxation oscillations (CSRO), 55
Community access television (CATV), 51
Compound semiconductor, 127, 128, 129, 139
Crystal orientation, 183

D

Deal-Grove model, 211
Deformation potential, 186, 188
Device channel, 228
Distributed feedback (DFB), 55

E

Electro luminescence, 126, 127, 130, 145, 147, 148
Electron collision, 74
Electronic circuit, 76

Encapsulation, 1, 2, 3, 4, 7, 8, 9, 10, 11, 12, 13, 15, 16, 17, 18, 23
Energy flux, 27, 30, 31

F

Fourier law closure, 26

G

Gate-dielectrics, 210

H

Halide transport method, 125
Hydrodynamic model, 26, 47, 48

I

Identity tensor, 27
Indium tin oxide (ITO), 7
Interface trap density, 245, 246, 247, 248, 252, 253, 257, 258, 262
Ionization, 74, 75, 77, 78, 79, 80, 83, 84, 86, 93, 96, 97, 113, 114, 115, 117, 118

L

Laser diode (LD), 125
Lithography, 228
Low-cost fabrication, 1

M

Mass spectrometer, 74
Metal-oxide-semiconductor (MOS), 246
Modeler geometry, 233
Modulation, 51, 54, 55, 56, 57, 59, 61, 64, 65, 67, 68, 69, 70, 71

N

Nano wire sub-band structure, 185
Nano-wire, 183, 184, 185, 186, 187, 188, 189, 190, 194, 195, 196, 198, 203, 204, 205, 206
Nanowire-based gate-all-around transistor, 184
National Television Standards Committee (NTSC), 51
Neighboring region, 234
Neutral oxidising, 209, 224
Normalization, 77
Numerical stability, 25

O

Optimal transport, 183
Organic electronics, 2
Organic light-emitting devices (OLEDs), 3
Organic material, 1, 2, 3, 10
Oscillator strength, 75, 76, 77, 91, 117, 118
Oxide layer, 210, 212
Oxygen permeation, 1, 2, 4
Oxygen transmission rate (OTR), 5

P

Passivation layer, 8
Passivation property, 246, 247, 258
Photo detector, 158, 159, 160, 179, 181, 182
Photo transistor, 157, 158, 160, 161, 162, 163, 164, 165, 166,

171, 172, 173, 174, 175, 176, 177, 178, 179, 180, 181, 182
Photodiode, 158, 159

Q

Quantization, 185, 187, 188, 195, 198, 200, 201
Quantum confinement transport, 25, 43, 45, 47, 48
Quantum drift diffusion (QDD), 26
Quantum energy transport (QET), 25
Quantum hydrodynamic (QHD), 25, 26

R

Radio Corporation of America (RCA), 127
Relaxation oscillation, 51, 55, 64

S

Scaling potential, 245, 246
Semiconductor lasers (SLs), 51
Sequential Color with Memory (SECAM), 52
Silicon allow, 158
Silicon dioxide, 209, 210, 223, 224
Silicon oxidation, 210, 211, 223, 224, 225
Silicon photodetector, 159
Silicon technology, 210
Silicon-on-insulator (SOI), 184
Simulation mesh, 234
Spontaneous emission, 54, 56, 58
Strain calculation, 230, 241
Strain engineering, 228
Strained silicon, 233, 234, 238
Sub-threshold, 184
Surface inversion layer, 232

T

Tetrahedral element, 233
Thermonuclear fusion, 73, 83, 86, 103, 115
Thin film encapsulation (TFE), 3
Thin-film encapsulation (TFE), 1
Thin-film encapsulation technology, 3
Transformation, 209

U

Ultra-thin strained silicon, 183

V

Valence band, 185
Vitex multilayer technology, 2

W

Wigner–Boltzmann equation, 25